LONDON MATHEMATICAL SOCIETY LECTURE NOTE SERIES

Managing Editor: Professor M. Reid, Mathematics Institute,
University of Warwick, Coventry CV4 7AL, United Kingdom

The titles below are available from booksellers, or from Cambridge University Press at
http://www.cambridge.org/mathematics

332 Handbook of tilting theory, L. ANGELERI HÜGEL, D. HAPPEL & H. KRAUSE (eds)
333 Synthetic differential geometry (2nd Edition), A. KOCK
334 The Navier–Stokes equations, N. RILEY & P. DRAZIN
335 Lectures on the combinatorics of free probability, A. NICA & R. SPEICHER
336 Integral closure of ideals, rings, and modules, I. SWANSON & C. HUNEKE
337 Methods in Banach space theory, J.M.F. CASTILLO & W.B. JOHNSON (eds)
338 Surveys in geometry and number theory, N. YOUNG (ed)
339 Groups St Andrews 2005 I, C.M. CAMPBELL, M.R. QUICK, E.F. ROBERTSON & G.C. SMITH (eds)
340 Groups St Andrews 2005 II, C.M. CAMPBELL, M.R. QUICK, E.F. ROBERTSON & G.C. SMITH (eds)
341 Ranks of elliptic curves and random matrix theory, J.B. CONREY, D.W. FARMER, F. MEZZADRI & N.C. SNAITH (eds)
342 Elliptic cohomology, H.R. MILLER & D.C. RAVENEL (eds)
343 Algebraic cycles and motives I, J. NAGEL & C. PETERS (eds)
344 Algebraic cycles and motives II, J. NAGEL & C. PETERS (eds)
345 Algebraic and analytic geometry, A. NEEMAN
346 Surveys in combinatorics 2007, A. HILTON & J. TALBOT (eds)
347 Surveys in contemporary mathematics, N. YOUNG & Y. CHOI (eds)
348 Transcendental dynamics and complex analysis, P.J. RIPPON & G.M. STALLARD (eds)
349 Model theory with applications to algebra and analysis I, Z. CHATZIDAKIS, D. MACPHERSON, A. PILLAY & A. WILKIE (eds)
350 Model theory with applications to algebra and analysis II, Z. CHATZIDAKIS, D. MACPHERSON, A. PILLAY & A. WILKIE (eds)
351 Finite von Neumann algebras and masas, A.M. SINCLAIR & R.R. SMITH
352 Number theory and polynomials, J. MCKEE & C. SMYTH (eds)
353 Trends in stochastic analysis, J. BLATH, P. MÖRTERS & M. SCHEUTZOW (eds)
354 Groups and analysis, K. TENT (ed)
355 Non-equilibrium statistical mechanics and turbulence, J. CARDY, G. FALKOVICH & K. GAWEDZKI
356 Elliptic curves and big Galois representations, D. DELBOURGO
357 Algebraic theory of differential equations, M.A.H. MACCALLUM & A.V. MIKHAILOV (eds)
358 Geometric and cohomological methods in group theory, M.R. BRIDSON, P.H. KROPHOLLER & I.J. LEARY (eds)
359 Moduli spaces and vector bundles, L. BRAMBILA-PAZ, S.B. BRADLOW, O. GARCÍA-PRADA & S. RAMANAN (eds)
360 Zariski geometries, B. ZILBER
361 Words: Notes on verbal width in groups, D. SEGAL
362 Differential tensor algebras and their module categories, R. BAUTISTA, L. SALMERÓN & R. ZUAZUA
363 Foundations of computational mathematics, Hong Kong 2008, F. CUCKER, A. PINKUS & M.J. TODD (eds)
364 Partial differential equations and fluid mechanics, J.C. ROBINSON & J.L. RODRIGO (eds)
365 Surveys in combinatorics 2009, S. HUCZYNSKA, J.D. MITCHELL & C.M. RONEY-DOUGAL (eds)
366 Highly oscillatory problems, B. ENGQUIST, A. FOKAS, E. HAIRER & A. ISERLES (eds)
367 Random matrices: High dimensional phenomena, G. BLOWER
368 Geometry of Riemann surfaces, F.P. GARDINER, G. GONZÁLEZ-DIEZ & C. KOUROUNIOTIS (eds)
369 Epidemics and rumours in complex networks, M. DRAIEF & L. MASSOULIÉ
370 Theory of p-adic distributions, S. ALBEVERIO, A.YU. KHRENNIKOV & V.M. SHELKOVICH
371 Conformal fractals, F. PRZYTYCKI & M. URBAŃSKI
372 Moonshine: The first quarter century and beyond, J. LEPOWSKY, J. MCKAY & M.P. TUITE (eds)
373 Smoothness, regularity and complete intersection, J. MAJADAS & A. G. RODICIO
374 Geometric analysis of hyperbolic differential equations: An introduction, S. ALINHAC
375 Triangulated categories, T. HOLM, P. JØRGENSEN & R. ROUQUIER (eds)
376 Permutation patterns, S. LINTON, N. RUŠKUC & V. VATTER (eds)
377 An introduction to Galois cohomology and its applications, G. BERHUY
378 Probability and mathematical genetics, N. H. BINGHAM & C. M. GOLDIE (eds)
379 Finite and algorithmic model theory, J. ESPARZA, C. MICHAUX & C. STEINHORN (eds)
380 Real and complex singularities, M. MANOEL, M.C. ROMERO FUSTER & C.T.C WALL (eds)
381 Symmetries and integrability of difference equations, D. LEVI, P. OLVER, Z. THOMOVA & P. WINTERNITZ (eds)
382 Forcing with random variables and proof complexity, J. KRAJÍČEK
383 Motivic integration and its interactions with model theory and non-Archimedean geometry I, R. CLUCKERS, J. NICAISE & J. SEBAG (eds)
384 Motivic integration and its interactions with model theory and non-Archimedean geometry II, R. CLUCKERS, J. NICAISE & J. SEBAG (eds)
385 Entropy of hidden Markov processes and connections to dynamical systems, B. MARCUS, K. PETERSEN & T. WEISSMAN (eds)

386	Independence-friendly logic, A.L. MANN, G. SANDU & M. SEVENSTER
387	Groups St Andrews 2009 in Bath I, C.M. CAMPBELL et al (eds)
388	Groups St Andrews 2009 in Bath II, C.M. CAMPBELL et al (eds)
389	Random fields on the sphere, D. MARINUCCI & G. PECCATI
390	Localization in periodic potentials, D.E. PELINOVSKY
391	Fusion systems in algebra and topology, M. ASCHBACHER, R. KESSAR & B. OLIVER
392	Surveys in combinatorics 2011, R. CHAPMAN (ed)
393	Non-abelian fundamental groups and Iwasawa theory, J. COATES et al (eds)
394	Variational problems in differential geometry, R. BIELAWSKI, K. HOUSTON & M. SPEIGHT (eds)
395	How groups grow, A. MANN
396	Arithmetic differential operators over the p-adic integers, C.C. RALPH & S.R. SIMANCA
397	Hyperbolic geometry and applications in quantum chaos and cosmology, J. BOLTE & F. STEINER (eds)
398	Mathematical models in contact mechanics, M. SOFONEA & A. MATEI
399	Circuit double cover of graphs, C.-Q. ZHANG
400	Dense sphere packings: a blueprint for formal proofs, T. HALES
401	A double Hall algebra approach to affine quantum Schur–Weyl theory, B. DENG, J. DU & Q. FU
402	Mathematical aspects of fluid mechanics, J.C. ROBINSON, J.L. RODRIGO & W. SADOWSKI (eds)
403	Foundations of computational mathematics, Budapest 2011, F. CUCKER, T. KRICK, A. PINKUS & A. SZANTO (eds)
404	Operator methods for boundary value problems, S. HASSI, H.S.V. DE SNOO & F.H. SZAFRANIEC (eds)
405	Torsors, étale homotopy and applications to rational points, A.N. SKOROBOGATOV (ed)
406	Appalachian set theory, J. CUMMINGS & E. SCHIMMERLING (eds)
407	The maximal subgroups of the low-dimensional finite classical groups, J.N. BRAY, D.F. HOLT & C.M. RONEY-DOUGAL
408	Complexity science: the Warwick master's course, R. BALL, V. KOLOKOLTSOV & R.S. MACKAY (eds)
409	Surveys in combinatorics 2013, S.R. BLACKBURN, S. GERKE & M. WILDON (eds)
410	Representation theory and harmonic analysis of wreath products of finite groups, T. CECCHERINI-SILBERSTEIN, F. SCARABOTTI & F. TOLLI
411	Moduli spaces, L. BRAMBILA-PAZ, O. GARCÍA-PRADA, P. NEWSTEAD & R.P. THOMAS (eds)
412	Automorphisms and equivalence relations in topological dynamics, D.B. ELLIS & R. ELLIS
413	Optimal transportation, Y. OLLIVIER, H. PAJOT & C. VILLANI (eds)
414	Automorphic forms and Galois representations I, F. DIAMOND, P.L. KASSAEI & M. KIM (eds)
415	Automorphic forms and Galois representations II, F. DIAMOND, P.L. KASSAEI & M. KIM (eds)
416	Reversibility in dynamics and group theory, A.G. O'FARRELL & I. SHORT
417	Recent advances in algebraic geometry, C.D. HACON, M. MUSTAȚĂ & M. POPA (eds)
418	The Bloch–Kato conjecture for the Riemann zeta function, J. COATES, A. RAGHURAM, A. SAIKIA & R. SUJATHA (eds)
419	The Cauchy problem for non-Lipschitz semi-linear parabolic partial differential equations, J.C. MEYER & D.J. NEEDHAM
420	Arithmetic and geometry, L. DIEULEFAIT et al (eds)
421	O-minimality and Diophantine geometry, G.O. JONES & A.J. WILKIE (eds)
422	Groups St Andrews 2013, C.M. CAMPBELL et al (eds)
423	Inequalities for graph eigenvalues, Z. STANIĆ
424	Surveys in combinatorics 2015, A. CZUMAJ et al (eds)
425	Geometry, topology and dynamics in negative curvature, C.S. ARAVINDA, F.T. FARRELL & J.-F. LAFONT (eds)
426	Lectures on the theory of water waves, T. BRIDGES, M. GROVES & D. NICHOLLS (eds)
427	Recent advances in Hodge theory, M. KERR & G. PEARLSTEIN (eds)
428	Geometry in a Fréchet context, C. T. J. DODSON, G. GALANIS & E. VASSILIOU
429	Sheaves and functions modulo p, L. TAELMAN
430	Recent progress in the theory of the Euler and Navier-Stokes equations, J.C. ROBINSON, J.L. RODRIGO, W. SADOWSKI & A. VIDAL-LÓPEZ (eds)
431	Harmonic and subharmonic function theory on the real hyperbolic ball, M. STOLL
432	Topics in graph automorphisms and reconstruction (2nd Edition), J. LAURI & R. SCAPELLATO
433	Regular and irregular holonomic D-modules, M. KASHIWARA & P. SCHAPIRA
434	Analytic semigroups and semilinear initial boundary value problems (2nd Edition), K. TAIRA
435	Graded rings and graded Grothendieck groups, R. HAZRAT
436	Groups, graphs and random walks, T. CECCHERINI-SILBERSTEIN, M. SALVATORI & E. SAVA-HUSS (eds)
437	Dynamics and analytic number theory, D. BADZIAHIN, A. GORODNIK & N. PEYERIMHOFF (eds)
438	Random walks and heat kernels on graphs, M.T. BARLOW
439	Evolution equations, K. AMMARI & S. GERBI (eds)
440	Surveys in combinatorics 2017, A. CLAESSON et al (eds)
441	Polynomials and the mod 2 Steenrod algebra I, G. WALKER & R.M.W. WOOD
442	Polynomials and the mod 2 Steenrod algebra II, G. WALKER & R.M.W. WOOD
443	Asymptotic analysis in general relativity, T. DAUDÉ, D. HÄFNER & J.-P. NICOLAS (eds)
444	Geometric and cohomological group theory, P.H. KROPHOLLER, I.J. LEARY, C. MARTÍNEZ-PÉREZ & B.E.A. NUCINKIS (eds)
445	Introduction to hidden semi-Markov models, J. VAN DER HOEK & R.J. ELLIOTT
446	Advances in two-dimensional homotopy and combinatorial group theory, W. METZLER & S. ROSEBROCK (eds)

London Mathematical Society Lecture Note Series: 446

Advances in Two-Dimensional Homotopy and Combinatorial Group Theory

Edited by

WOLFGANG METZLER
Johann Wolfgang Goethe-Universität Frankfurt, Germany

STEPHAN ROSEBROCK
Pädagogische Hochschule Karlsruhe, Germany

CAMBRIDGE
UNIVERSITY PRESS

University Printing House, Cambridge CB2 8BS, United Kingdom

One Liberty Plaza, 20th Floor, New York, NY 10006, USA

477 Williamstown Road, Port Melbourne, VIC 3207, Australia

314-321, 3rd Floor, Plot 3, Splendor Forum, Jasola District Centre, New Delhi - 110025, India

79 Anson Road, #06-04/06, Singapore 079906

Cambridge University Press is part of the University of Cambridge.

It furthers the University's mission by disseminating knowledge in the pursuit of education, learning and research at the highest international levels of excellence.

www.cambridge.org
Information on this title: www.cambridge.org/9781316600900
DOI: 10.1017/9781316555798

© Cambridge University Press 2018

This publication is in copyright. Subject to statutory exception and to the provisions of relevant collective licensing agreements, no reproduction of any part may take place without the written permission of Cambridge University Press.

First published 2018

A catalogue record for this publication is available from the British Library

ISBN 978-1-316-60090-0 Paperback

Additional resources for this publication at www.cambridge.org/9781316600900

Cambridge University Press has no responsibility for the persistence or accuracy of URLs for external or third-party internet websites referred to in this publication, and does not guarantee that any content on such websites is, or will remain, accurate or appropriate.

Contents

Editors' Preface		*page* ix
Addresses of Authors		xi
1	**A Survey of recent Progress on some Problems in 2-dimensional Topology**	
	Jens Harlander	1
1.1	Introduction	1
1.2	Basic definitions	2
1.3	Basic questions	7
1.4	The D(2)-problem and the Eilenberg-Ganea conjecture	9
1.5	Dunwoody's counterexample to relation lifting	11
1.6	Solving the realization problem for some groups	13
1.7	Exotic presentations of the trefoil group	15
1.8	Exotic almost presentations for the Klein bottle group	17
1.9	Appendix: Geometric Realization for Algebraic 2-Complexes with Finite Fundamental Group *By F. Rudolf Beyl and Jens Harlander*	19
2	**Further Results concerning the Andrews-Curtis-Conjecture and its Generalizations**	
	Cynthia Hog-Angeloni and Wolfgang Metzler	27
2.1	Introduction	27
2.2	Projections onto test groups	28
2.3	Fixed subcomplexes	30
2.4	(AC) for spines of 3-manifolds	33
2.5	Considerations on length	34
2.6	On the future of the Andrews-Curtis problem	35

3 Aspects of TQFT and Computational Algebra
Holger Kaden and Simon King   36
   3.1 Introduction   36
   3.2 The Concept of Topological Quantum Field theory (TQFT)   37
   3.3 TQFT on 2-complexes   40
   3.4 TQFT on s-move 3-cells   52
   3.5 A general construction of state sum invariants   58
   3.6 Matveev-Piergallini calculus   68

4 Labelled Oriented Trees and the Whitehead-Conjecture
Stephan Rosebrock   72
   4.1 Introduction   72
   4.2 The finite and the infinite case   73
   4.3 Spherical diagrams   74
   4.4 Classes of aspherical LOTs   77
   4.5 The asphericity of injective LOTs   86
   4.6 Two examples   95
   4.7 Virtual Knots and LOTs   96
   4.8 L^2-homology and Whitehead's asphericity conjecture
By Jens Harlander   98

5 2-Complexes and 3-Manifolds
Janina Glock, Cynthia Hog-Angeloni and Sergei Matveev   103
   5.1 Introduction   103
   5.2 Existence and Uniqueness   104
   5.3 k-Connectivity of Whitehead Graphs   106
   5.4 M.H.A. Newman's Diamond Lemma and a new version of it   114
   5.5 Universal Scheme   117
   5.6 Kneser-Milnor Theorem   118
   5.7 Prime decompositions of knots in $F \times I$ 120

6 The Relation Gap Problem
Jens Harlander   128
   6.1 Introduction   128
   6.2 The Relation Gap problem   129
   6.3 Examples   132
   6.4 An Infinite Relation Gap   133
   6.5 Efficiency   137
   6.6 Presentations with Cyclic Relation Modules   142
   6.7 Finite Groups   143

7 On the Relation Gap Problem for Free Products
Cynthia Hog-Angeloni and Wolfgang Metzler   149
   7.1   Introduction   149
   7.2   Semisplit presentations, flower relations and integral flower chains   150
   7.3   Applications to the Epstein Groups   158
   7.4   Ideas for further work   159
   7.5   Flower chains and Splitting with respect to field coefficients   161
   7.6   Proof of the Theorem on semisplit Q^{**}-transformations   165

References   167
Index   177
Erratum   180

Editors' Preface

Our book on "Two-dimensional Homotopy and Combinatorial Group Theory (LMS 197)" appeared 24 years ago. Since the turn of the millennium we were encouraged to publish in a similar style the (remarkable) progress that has been obtained in the meantime. On the other hand, none of the fundamental open questions like the Andrews-Curtis Conjecture and the Whitehead-Asphericity problem have been (fully) solved, and almost no prognosis is possible about when it will finally happen. We hope that mathematicians, who are in danger of leaving the subject because of this perspective, will be kept active by this sequel publication; and new ones will be gained for our important field.

Again, we mix surveys and original material in the chapters, which can be read by students and experts alike. Ongoing research is mentioned as work in progress. There are ample references to the LMS 197 volume which will be cited as [I] including the bibliography and the index, where new publications and new notions are listed.

Since the publication of LMS 197 there have been workshops of the authors at least every fifth year. We met in Germany, Italy, Russia and the USA, and the fruits of these meetings are the basis of this book. During a meeting close to Naumburg (Saale) we made the decision to start what you can read now.

A mathematically energetic member of the team of authors was Nancy Waller. But when the actual writing had to start, she unexpectedly died. Her preprint [BeWa13] together with F. Rudolf Beyl could not be finished, and in particular it was impossible without her to write a planned chapter on the *geometric realization problem for algebraic 2-complexes* at the same time. Hence "Nancy's Toy", which she hoped that would result in a counterexample, has been left to the future. What can be said at present, is collected in an appendix to chapter I by F. Rudolf Beyl and Jens Harlander. This book can be considered as serving for Nancy Waller's memory.

As for the LMS 197 volume, all chapters are refereed within the team of authors and by an external referee who had the option of remaining anonymous. We are grateful for their valuable service. Those who gave us permission to be named openly are Laura Biroth (Mainz), Martin Bridson (Oxford), Cameron Gordon (Austin, TX), Paul Latiolais (Portland, OR), Peter Linnell (Blacksburg, VA), Wajid Mannan (Brighton), and Timo deWolff (College Station, TX). We also thank our former students for their help to prepare this book, some of whom are mentioned in the chapters and in the bibliography. A special thank you goes to Wesley Browning, who allowed us to reproduce his note [Bro76], which grew out of correspondence with Joan Birman and which should not be forgotten.

Last, but not least, we again thank Roger Astley and his team, who gave encouragement and continuous advice as for LMS 197.

July 2017
Wolfgang Metzler and Stephan Rosebrock

Addresses of Authors

F. Rudolf Beyl
Fariborz Maseeh Department of
Mathematics
Portland State University
Portland, OR 97201/USA

Janina Glock
Offenbacher Str. 32
63165 Mühlheim/Germany

Jens Harlander
Department of Mathematics
Boise State University
1910 University Drive
Boise, ID 83725-1555, USA

Cynthia Hog-Angeloni
Johannes-Gutenberg-Universität Mainz
Institut für Mathematik
Staudingerweg 9
55099 Mainz/Germany

Holger Kaden
Theodor-Heuss-Str. 2
61130 Nidderau/Germany

Simon King
Fakultät für Mathematik und Informatik
Friedrich-Schiller-Universität Jena
07737 Jena/Germany

Sergei Matveev
Chelyabinsk State University
Br. Kashirinykh Str. 129
454001 Chelyabinsk/Russia

Wolfgang Metzler
Johann Wolfgang Goethe-Universität
FB 12 Institut für Mathematik
Robert-Mayer-Str. 6-8
60325 Frankfurt am Main/Germany

Stephan Rosebrock
Pädagogische Hochschule Karlsruhe
Bismarckstraße 10
76133 Karlsruhe /Germany

1
A Survey of recent Progress on some Problems in 2-dimensional Topology

Jens Harlander

1.1 Introduction

A primary theme in algebraic topology is the characterization of the properties of a space or group in homological terms. Specific examples are Wall's questions concerning finiteness properties of spaces, or the Eilenberg-Ganea problem that compares the cohomological dimension of a group with its geometric dimension. Relevant to both examples is the question: can a partial resolution of length n be geometrically realized?

Any partial resolution of length n is chain homotopically equivalent to one of length n with geometric $(n-1)$-skeleton. For $n \geq 3$, the $(n-1)$-skeleton of the universal covering of a classifying space is simply connected, and this fact can be used to obtain a geometric realizations of the original partial resolution. For $n = 2$, however, simple connectivity fails and the realization question translates into difficult questions concerning presentations of groups. These questions are the topic of this chapter.

We will introduce the *geometric realization problem* in dimension 2, the *relation lifting problem* and the *relation gap problem*. The first in this list of problems is the most fundamental, while the second and third can be stated purely in terms of combinatorial group theory.

The final section in this chapter is an appendix jointly written by F. Rudolf Beyl and the author. It gives a brief overview of the state of the geometric realization problem for finite groups.

This first chapter also serves as an introduction to two later chapters: Chapter 6 is a detailed account of the relation gap problem, and Chapter 7 discusses the relation gap problem in the context of free products.

1.2 Basic definitions

Let G be a group and **g** be a generating set for G. Let **x** be a set of letters and $\phi: \mathbf{x} \to \mathbf{g}$ be a bijection. This bijection extends to a group epimorphism

$$\phi: F(\mathbf{x}) \longrightarrow G,$$

from the free group $F = F(\mathbf{x})$ onto G. The kernel N of ϕ is called the *relation group* associated with the generating set **g** of G. The conjugation action of F on N provides a $\mathbb{Z}G$-module structure on $N_{ab} = N/[N,N]$: if $g \in G$ and $r \in N$, then the action of G on $N/[N,N]$ is well-defined by $g \cdot r[N,N] = frf^{-1}[N,N]$, with any $f \in F$ such that $g = \phi(f)$. This module is called the *relation module* associated with the generating set **g** of G. If we choose a set **r** of normal generators of N, then

$$P = \langle \mathbf{x} \mid \mathbf{r} \rangle$$

is called a *presentation* of the group G. By a slight abuse of notation we also refer to

$$1 \longrightarrow N \longrightarrow F(\mathbf{x}) \stackrel{\phi}{\longrightarrow} G \longrightarrow 1,$$

or just the quotient F/N, as a presentation of the group G. A presentation is called *finite* if both **x** and **r** are finite sets.

The *standard 2-complex* $K(P)$ associated with P is constructed as follows: $K(P)$ has a single 0-cell, 1-cells in one-to-one correspondence with the generators in P, and 2-cells in one-to-one correspondence with the relators $r \in \mathbf{r}$ in P. A word in the generators and their inverses defines a path in $K(P)$. The 2-cell in $K(P)$ corresponding to a relator r is attached to the 1-skeleton via the path corresponding to the word r in the generators. For convenience we will simply refer to the x_i as 1-cells of $K(P)$ and the $r \in \mathbf{r}$ as 2-cells of $K(P)$.

We will next describe the universal covering $\tilde{K}(P)$ of $K(P)$. The complex $\tilde{K}(P)$ has 0-cells in one-to-one correspondence with the elements of G, and oriented edges $g\tilde{x}$, $g \in G$, $x \in \mathbf{x}$. The initial vertex of $g\tilde{x}$ is g, its terminal vertex is $g\phi(x)$. We call edges of the form $g\tilde{x}$ edges with label x. Note that at a vertex g we have a unique outgoing edge with label x (the edge $g\tilde{x}$) and a unique incoming edge with label x (the edge $g\phi(x)^{-1}\tilde{x}$). Thus a path in $K(P)$, which corresponds to a word w in the generators and their inverses, can be uniquely lifted to a path $g\tilde{w}$, starting at g and ending at $g\phi(w)$. Reading the labels along the path $g\tilde{w}$ reads the word w. For every $g \in G$ and $r \in \mathbf{r}$ we have a 2-cell in $\tilde{K}(P)$ attached along the closed path $g\tilde{r}$. For convenience we will simply refer to $g\tilde{r}$ as a 2-cell in $\tilde{K}(P)$.

The 1-skeleton $\tilde{K}(P)^{(1)}$ is also called the *Cayley graph* $\Gamma(G, \mathbf{g})$ of G associated with the generating set $\mathbf{g} = \phi(\mathbf{x})$. A word w in $\mathbf{x}^{\pm 1}$ lifts to a closed path $\tilde{w}$ in $\Gamma(G, \mathbf{g})$, starting and ending at 1, if and only if $w \in N$. This defines a group isomorphism

$$N \longrightarrow \pi_1(\Gamma(G, \mathbf{g}), 1)$$

and a $\mathbb{Z}G$-module isomorphism

$$N/[N, N] \longrightarrow H_1(\Gamma(G, \mathbf{g})).$$

We next discuss the augmented cellular chain complex for $\tilde{K}(P)$, denoted by $C(P)$:

$$C(P) = (C_2(\tilde{K}(P)) \xrightarrow{\partial_2} C_1(\tilde{K}(P)) \xrightarrow{\partial_1} C_0(\tilde{K}(P)) \xrightarrow{\epsilon} \mathbb{Z} \longrightarrow 0).$$

Standard $\mathbb{Z}G$-module bases for $C_2(\tilde{K}(P))$, $C_1(\tilde{K}(P))$, and $C_0(\tilde{K}(P))$ are $\{\tilde{r} \mid r \in \mathbf{r}\}$, $\{\tilde{x} \mid x \in \mathbf{x}\}$, and $\{1\}$, respectively. We have

$$\partial_2(\tilde{r}) = \sum_{x \in \mathbf{x}} \phi(\frac{\partial r}{\partial x})\tilde{x},$$

$$\partial_1(\tilde{x}) = \phi(x) - 1,$$

$$\epsilon(1) = 1.$$

The coefficients $\frac{\partial r}{\partial x} \in \mathbb{Z}F(\mathbf{x})$ are called the *Fox derivatives* of r with respect to x. The epimorphism $\phi \colon F(\mathbf{x}) \to G$ induces a map on the respective group rings which we also denote by ϕ. For more details on the Fox derivatives see Lyndon/Schupp [I], [LySc77], Chapter I Section 10 and Chapter II Section 3, or Sieradski's Chapter II in [I]. Since the 1-skeleton of $\tilde{K}(P)$ is the Cayley graph $\Gamma(G, \mathbf{g})$, the kernel of ∂_1 is $H_1(\Gamma(G, \mathbf{g}))$. And since the relation module $N/[N, N]$ is isomorphic to that homology group, we have an exact sequence

$$0 \longrightarrow N/[N, N] \xrightarrow{\partial} C_1(\Gamma(G, \mathbf{g})) \xrightarrow{\partial_1} C_0(\Gamma(G, \mathbf{g})) \xrightarrow{\epsilon} \mathbb{Z} \longrightarrow 0, \qquad (1.1)$$

where

$$\partial(r[N, N]) = \sum_{x \in \mathbf{x}} \phi(\frac{\partial r}{\partial x})\tilde{x}.$$

We refer to this sequence as the *4-term exact sequence* associated with the pair $(G, \mathbf{g})$.

A $(G, 2)$-*complex* K is a connected 2-dimensional CW-complex with fundamental group G. An *algebraic* $(G, 2)$-*complex* A is an exact sequence

$$A = (A_2 \xrightarrow{\partial_2} A_1 \xrightarrow{\partial_1} A_0 \xrightarrow{\epsilon} \mathbb{Z} \longrightarrow 0), \tag{1.2}$$

where the A_i are free $\mathbb{Z}G$-modules. We say the complex is *finite* if the A_i are finitely generated.

The 4-term exact sequence (1.1) can be used to construct algebraic $(G, 2)$-complexes. Given

$$1 \longrightarrow N \longrightarrow F(\mathbf{x}) \xrightarrow{\phi} G \longrightarrow 1$$

we can choose subsets $\mathbf{s} \subseteq N$ and $\mathbf{c} \subseteq [N, N]$ so that the set $\{s[N, N] \mid s \in \mathbf{s}\}$ generates the relation module and the union $\mathbf{s} \cup \mathbf{c}$ normally generates N. So

$$P = \langle \mathbf{x} \mid \mathbf{s} \cup \mathbf{c} \rangle$$

is a presentation of G. We call

$$P_0 = [\mathbf{x} \mid \mathbf{s}]$$

an *almost presentation* of G. We have

$$C_2(P) = C_2(\tilde{K}(P)) = A_2 \oplus B_2,$$

where A_2 is a free $\mathbb{Z}G$-module with basis $\{\tilde{s} \mid s \in \mathbf{s}\}$ and B_2 is a free $\mathbb{Z}G$-module with basis $\{\tilde{c} \mid c \in \mathbf{c}\}$. Note that $\partial_2(B_2) = 0$. The algebraic $(G, 2)$-complex $A(P_0)$ is obtained from $C(P)$ by removing the factor B_2 from $C_2(P)$. So

$$A(P_0) = (A_2 \xrightarrow{\partial_2} C_1(\Gamma(G, \mathbf{g})) \xrightarrow{\partial_1} C_0(\Gamma(G, \mathbf{g})) \xrightarrow{\epsilon} \mathbb{Z} \longrightarrow 0),$$

where A_2 is free with basis $\{\tilde{s} \mid s \in \mathbf{s}\}$ and

$$\partial_2(\tilde{s}) = \sum_{x \in \mathbf{x}} \phi(\frac{\partial s}{\partial x})\tilde{x}.$$

If G is assumed to be finitely presented then the presentation P can be taken to be finite and hence $A(P_0)$ is a finite algebraic $(G, 2)$-complex. The complexes $C(P)$ and $A(P_0)$ are the objects of main interest in this chapter.

Theorem 1.1 *Let $\mathbf{s}$ be a subset of $F(\mathbf{x})$ and let J be the normal subgroup normally generated by $\mathbf{s}$. Let $\hat{G} = F(\mathbf{x})/J$. Then $P_0 = [\mathbf{x} \mid \mathbf{s}]$ is an almost presentation of G if and only if there exists a short exact sequence*

$$1 \longrightarrow H = N/J \longrightarrow \hat{G} = F(\mathbf{x})/J \longrightarrow G \longrightarrow 1$$

with $H_{ab} = 0$.

Proof: If P_0 is an almost presentation of G, then there exists a normal subgroup N of $F(\mathbf{x})$ and a subset $\mathbf{c} \subseteq [N, N]$ such that N is normally generated by $\mathbf{s} \cup \mathbf{c}$ and F/N is isomorphic to G. It follows that $N = [N, N]J$, and thus $H_{ab} = N/[N, N]J = 0$.

On the other hand, suppose we have a short exact sequence as given in the statement of the lemma. Without loss of generality we may take N to be the kernel of the composition

$$F(\mathbf{x}) \longrightarrow F(\mathbf{x})/J \longrightarrow G \longrightarrow 1.$$

Then $H_{ab} = 0$ implies that $N = [N, N]J$. Thus we can find a subset $\mathbf{c} \subseteq [N, N]$ such that N is normally generated by $\mathbf{s} \cup \mathbf{c}$, and so $P_0 = [\mathbf{x} \mid \mathbf{s}]$ is an almost presentation of G. □

Using the above theorem we can give another way to arrive at $A(P_0)$. Let $\hat{P} = \langle \mathbf{x} \mid \mathbf{s} \rangle$ be the presentation obtained from P_0 and let $\hat{G}$ be the group presented by $\hat{P}$. By Theorem 1.1 we have an epimorphism $\hat{G} \to G$ with perfect kernel H, that is $H_{ab} = 0$. Now $C(\hat{P})$ is an algebraic $(\hat{G}, 2)$-complex on which H acts, so we can form the quotient complex $\mathbb{Z} \otimes_{\mathbb{Z}H} C(\hat{P})$. Note that this quotient complex is an algebraic $(G, 2)$-complex because its first homology is the first homology of H and hence is trivial. Thus we have

$$A(P_0) = \mathbb{Z} \otimes_{\mathbb{Z}H} C(\hat{P}).$$

We can formulate this in geometric terms: If $\bar{K}$ is the covering of $K(\hat{P})$ associated with H, then $A(P_0) = C(\bar{K})$.

Theorem 1.2 *If A is an algebraic $(G, 2)$-complex, then A is chain-homotopically equivalent to some $A(P_0)$, where P_0 is an almost presentation of G.*

Proof: The result follows from Theorem 1.1 in Mannan [Man07b]. We will provide some details.

Let

$$A = (A_2 \xrightarrow{\partial_2^A} A_1 \xrightarrow{\partial_1^A} A_0 \xrightarrow{\epsilon^A} \mathbb{Z} \longrightarrow 0).$$

Let $P = \langle \mathbf{x} \mid \mathbf{r} \rangle$ be a presentation for G and

$$C(P) = (C_2 \xrightarrow{\partial_2^C} C_1 \xrightarrow{\partial_1^C} C_0 \xrightarrow{\epsilon^C} \mathbb{Z} \longrightarrow 0)$$

be the associated algebraic 2-complex. By Theorem 1.1 in Mannan [Man07b] there exist free $\mathbb{Z}G$-modules S and T, such that

$$A'^{(1)} = (A_1 \oplus S \xrightarrow{\partial_1^A \oplus 0} A_0 \xrightarrow{\epsilon^A} \mathbb{Z} \longrightarrow 0)$$

and
$$C'^{(1)} = (C_1 \oplus T \xrightarrow{\partial_1^C \oplus 0} C_0 \xrightarrow{\epsilon^C} \mathbb{Z} \longrightarrow 0)$$
are chain homotopy equivalent. Let $\phi \colon A'^{(1)} \longrightarrow C'^{(1)}$ be a chain-homotopy equivalence. It induces an isomorphism
$$\phi \colon \ker \partial_1^A \oplus S \longrightarrow \ker \partial_1^C \oplus T.$$
We can construct the algebraic 2-complex
$$A' = (A_2 \oplus S \xrightarrow{\partial_2} C_1 \oplus T \xrightarrow{\partial_1^C \oplus 0} C_0 \xrightarrow{\epsilon^C} \mathbb{Z} \longrightarrow 0),$$
where
$$\partial_2 = \phi \circ (\partial_2^A \oplus 1).$$
Note that A' is chain-homotopically equivalent to
$$A'' = (A_2 \oplus S \xrightarrow{\partial_2^A \oplus 1} A_1 \oplus S \xrightarrow{\partial_1^A \oplus 0} A_0 \xrightarrow{\epsilon^A} \mathbb{Z} \longrightarrow 0).$$
Since A'' is clearly chain-homotopically equivalent to A, we conclude that A' is chain-homotopically equivalent to A. It remains to show that $A' = A(P_0)$ for some almost presentation P_0 of G. Let $\mathbf{y}$ be a set in one-to-one correspondence with a basis of T. Consider the presentation $P' = \langle \mathbf{x} \cup \mathbf{y} \mid \mathbf{r} \cup \mathbf{y} \rangle$ of G. Note that the 1-skeleton of $C(P')$ is the same as the 1-skeleton of A'. Thus
$$A' = (A'_2 \xrightarrow{\partial_2} C_1(\Gamma(G, \mathbf{g}')) \xrightarrow{\partial_1} C_0(\Gamma(G, \mathbf{g}')) \xrightarrow{\epsilon} \mathbb{Z} \longrightarrow 0),$$
where $A'_2 = A_2 \oplus S$ and $\mathbf{g}'$ is the generating set of G coming from $\mathbf{x} \cup \mathbf{y}$. It follows that $A' = A(P_0)$ for some almost presentation P_0 of G. □

The *Euler characteristic* of a finite algebraic $(G, 2)$-complex A is defined as
$$\chi(A) = m_0 - m_1 + m_2,$$
where m_i is the rank of the free module A_i, $i = 0, 1, 2$. Note than if K is a finite $(G, 2)$-complex then the chain complex of the universal covering $C(\tilde{K})$ is a finite algebraic $(G, 2)$-complex and $\chi(K) = \chi(C(\tilde{K}))$. For finitely presented groups G we define
$$\chi_{geom}(G, 2) = min\{\chi(K) \mid K \text{ is a finite } (G, 2)\text{-complex}\}$$
$$\chi_{alg}(G, 2) = min\{\chi(A) \mid A \text{ is a finite algebraic } (G, 2)\text{-complex}\}.$$
We have that
$$\chi_{alg}(G, 2) \leq \chi_{geom}(G, 2).$$

1.3 Basic questions

We say an algebraic $(G, 2)$-complex A is *geometrically realizable* if there exists a $(G, 2)$-complex K so that $C(\tilde{K})$ and A are chain-homotopically equivalent.

Geometric realization problem: Is every algebraic $(G, 2)$-complex geometrically realizable?

Using Theorem 1.2 and the fact that a $(G, 2)$-complex K is homotopically equivalent to a standard 2-complex $K(P)$, for some presentation P of G, we can state the geometric realization problem also in the following way: Given an almost presentation P_0 of G, does there exist a presentation P of G so that the algebraic 2-complexes $A(P_0)$ and $C(P)$ are chain-homotopically equivalent?

Bestvina and Brady [BeBr97] constructed a finite almost presentation P_0 for a certain group G which is not finitely presented. Thus $A(P_0)$ is not geometrically realizable. The geometric realization problem is open for finitely presented groups. More details on the Bestvina/Brady construction are provided in Chapter 6.

We will next state some well known problems from combinatorial group theory and frame them in terms of geometric realization.

Relation lifting problem: Let F/N be a finite presentation of the group G. Given a set $\mathbf{s} = \{s_1, ..., s_m\} \subseteq N$ so that the set $\{s_1[N, N], ..., s_m[N, N]\}$ generates the relation module; does there exist a set $\mathbf{r} = \{r_1, ..., r_m\}$ of normal generators of N, such that $r_i[N, N] = s_i[N, N], i = 1, ...m$?

We refer to the elements r_i as *lifts* of the generators $s_i[N, N], i = 1, ..., m$. In [Wall66] Wall conjectured that relation lifting is always possible. However, in [Dun72] Dunwoody provided an example where lifting is not possible. He utilized the existence of non-trivial units in the group ring of a certain group with torsion. For torsion-free groups the relation lifting problem is open. We will give more details on Dunwoody's construction in Section 1.5. The following theorem states the relation lifting problem in terms of geometric realization.

Theorem 1.3 *Let $P_0 = [\mathbf{x} \mid \mathbf{s}]$ be a finite almost presentation of G. Let N be the kernel of the composition $F(\mathbf{x}) \to \langle \mathbf{x} \mid \mathbf{s} \rangle \to G$. Then the relation module generators $\{s[N, N] \mid s \in \mathbf{s}\}$ can be lifted if and only if there exists a finite presentation P of G such that $C(P) = A(P_0)$.*

Proof: The relation module generators can be lifted if and only if there exist elements $c_s \in [N,N]$, $s \in \mathbf{s}$, so that $P = \langle \mathbf{x} \mid \mathbf{r} \rangle$, $\mathbf{r} = \{r_s = sc_s \mid s \in \mathbf{s}\}$, is a presentation of G. The chain complexes $C(P)$ and $A(P_0)$ agree in dimension 0 and 1. Now $C_2(P)$ free $\mathbb{Z}G$-module with basis $\{\tilde{r}_s \mid s \in \mathbf{s}\}$, and $A_2(P_0)$ is a free $\mathbb{Z}G$-module with basis $\{\tilde{s} \mid s \in \mathbf{s}\}$. Furthermore the boundary of $\tilde{s}$ agrees with the boundary of $\tilde{r}_s$. Thus the two chain complexes can be identified. The other direction follows from the 4-term exact sequence (1.1) given in Section 1.2. □

It follows that if $P_0 = [\mathbf{x} \mid \mathbf{s}]$ is an almost presentation of G and the relation module generators $\mathbf{s}[N,N]$ can not be lifted, then there is a chance that the algebraic $(G,2)$-complex $A(P_0)$ might not be geometrically realizable.

A fundamental problem in combinatorial group theory is to determine the minimal number of relations required to present a group on a given set of generators. Given a finite presentation F/N of a group G we have the chain of inequalities

$$d_F(N) \geq d_G(N/[N,N]) \geq d(N/[F,N]) = d(F) - \text{tfr}(H_1(G)) + d(H_2(G)).$$

Here $d_F(-)$ denotes the minimal number of normal generators, $d_G(-)$ denotes the minimal number of G-module generators, $d(-)$ denotes the minimal number of generators, and $\text{tfr}(-)$ denotes the torsion free rank. The chain of inequalities follows from the exact sequence

$$H_2(G) \longrightarrow H_2(Q) \longrightarrow \mathbb{Z} \otimes_{\mathbb{Z}G} H_1(H) \longrightarrow H_1(G) \longrightarrow H_1(Q) \longrightarrow 0,$$

associated with an exact sequence of groups $1 \to H \to G \to Q \to 1$ (see Brown [I], [Br82]). We say the presentation F/N has a relation gap if the first inequality in the above chain is strict:

$$d_F(N) > d_G(N/[N,N]).$$

Relation gap problem: Does there exist a finite presentation F/N with a relation gap?

Infinite relation gaps can occur. See Bestvina and Brady [BeBr97]. The relation gap problem is open for finitely presented groups. Chapter 6 in this book is devoted to this problem.

The next theorem relates the relation gap problem to the geometric realization problem.

Theorem 1.4 *Suppose there exists a finite presentation F/N of G so that*

1 $1 - d(F) + d_F(N) = \chi_{geom}(G, 2)$;
2 F/N *has a relation gap,*

Then there exists a finite algebraic $(G, 2)$-complex that can not be geometrically realized.

Proof: The conditions imply that $\chi_{alg}(G, 2) < \chi_{geom}(G, 2)$. □

1.4 The D(2)-problem and the Eilenberg-Ganea conjecture

In this section we will state two more well known problems and relate them to the questions discussed in the last section.

Let X be a connected CW-complex with fundamental group G and let M be a $\mathbb{Z}G$-module. Then $H^k(X, M)$ is defined as the kth-cohomology of $\tilde{X}$ with coefficients in M. We say that X has *cohomological dimension n* if there exists a $\mathbb{Z}G$-module M_0 so that $H^n(X, M_0) \neq 0$, but $H^k(X, M) = 0$ for $k > n$ and all $\mathbb{Z}G$-modules M.

In [I], [Wa65] Wall compared cohomological dimension and dimension of CW-complexes.

Theorem 1.5 *(Wall) Let X be a finite connected $(n + 1)$-dimensional CW-complex of cohomological dimension $n > 2$. If $H_{n+1}(\tilde{X}, \mathbb{Z}) = 0$, then X is homotopy equivalent to a finite n-dimensional CW-complex.*

The statement is obviously true for $n = 0$ and holds for $n = 1$ by the work of Stallings [Sta68] and Swan [Swa69]. Whether this theorem is true in case $n = 2$ is unknown. This case is known as the *D(2)-problem*.

D(2)-problem: Let X be a finite connected 3 dimensional complex of cohomological dimension 2. Assume that $H_3(\tilde{X}, \mathbb{Z}) = 0$. Is X homotopy equivalent to a finite 2-dimensional complex?

Remark. One can show (see Wall [I], [Wa65]) that the conditions given in the D(2)-problem are equivalent to stating that X is *dominated* by a finite 2-complex.

The connection between the D(2)-problem and the geometric realization problem was known to Wall (see [Wall66], Theorem 4). The equivalence of the two problems was made explicit by Johnson [Joh03] and Mannan [Man09].

Theorem 1.6 *The geometric realization problem for finitely presented groups and the D(2)-problem are equivalent.*

Let G be a group. We say that G has *cohomological dimension* n if there exists a $\mathbb{Z}G$-module M_0 so that $H^n(G, M_0) \neq 0$, but $H^k(G, M) = 0$ for $k > n$ and all $\mathbb{Z}G$-modules M. A connected complex X is called a $K(G, 1)$-complex if $\pi_1(X) = G$ and $\pi_n(X) = 0$ for $n > 1$. It is well known that all $K(G, 1)$-complexes are homotopically equivalent. The group is said to have *geometric dimension* n if there exists an n-dimensional $K(G, 1)$-complex, but none of smaller dimension. The following was shown by Eilenberg and Ganea [EiGa57].

Theorem 1.7 *(Eilenberg-Ganea) If $n > 2$ then G has cohomological dimension n if and only if G has geometric dimension n.*

As for Theorem 1.5 the statement is obviously true for $n = 0$ and holds for $n = 1$ by the work of Stallings [Sta68] and Swan [Swa69]. The case $n = 2$ is unknown.

Eilenberg-Ganea conjecture. The group G has cohomological dimension 2 if and only if G has geometric dimension 2.

It is well known ([I], [Br82]) that G has cohomological dimension n if and only if there exists a projective resolution

$$P = (0 \longrightarrow P_n \longrightarrow P_{n-1} \longrightarrow \ldots \longrightarrow P_0 \longrightarrow \mathbb{Z} \longrightarrow 0)$$

of length n, but none of shorter length. If G has geometric dimension 2 then there exists a 2-dimensional $K(G, 1)$-complex X and the chain complex $C(\tilde{X})$ provides a projective (even free) resolution of length 2. Hence only one direction in the conjecture is problematic, and it can be phrased as a geometric realization problem.

Reformulated Eilenberg-Ganea conjecture. Given a projective resolution P of $\mathbb{Z}G$-modules of length 2, does there exist a 2-complex X such that $C(\tilde{X})$ and P are chain-homotopically equivalent?

If P can be geometrically realized, then there exists a 2-complex X so that $C(\tilde{X})$ is homotopically equivalent to P. In particular $0 = H_2(P) = H_2(C(\tilde{X})) = \pi_2(X)$. Hence X is a finite 2-dimensional $K(G, 1)$-complex.

A restricted version of the Eilenberg-Ganea conjecture can be thought of as a special case of the geometric realization problem stated in Section 1.3.

Special Eilenberg-Ganea conjecture. Can a finite algebraic 2-complex

$$A = (A_2 \xrightarrow{\partial_2} A_1 \xrightarrow{\partial_1} A_0 \xrightarrow{\epsilon} \mathbb{Z} \longrightarrow 0),$$

where ∂_2 is injective, be geometrically realized?

1.5 Dunwoody's counterexample to relation lifting

In this section we present Dunwoody's counterexample to relation lifting in detail. It appeared in [Dun72]. Consider the presentation F/N where $F = F(a, b)$ is free of rank 2 and N is the normal closure of a^5. The quotient F/N presents the free product $G = \mathbb{Z}_5 * \mathbb{Z}$. In the following we think of elements of G as reduced normal form words in $\{a, b\}$. Note that $(1 - a + a^2)(a + a^2 - a^4) = 1$ in $\mathbb{Z}G$, so $1 - a + a^2$ is a non-trivial unit in $\mathbb{Z}G$. Hence so is

$$\alpha = (1 - a + a^2)b = b - ab + a^2 b.$$

It follows that

$$\alpha \cdot a^5[N, N] = (ba^5 b^{-1})(aba^{-5}b^{-1}a^{-1})(a^2 ba^5 b^{-1} a^{-2})[N, N]$$

is a generator of the relation module $N/[N, N]$. Let s be the word in $\{a, b\}^{\pm 1}$ that occurs on the right hand side of the equation. We will see that $s[N, N]$ cannot be lifted. For suppose that r is a lift. Then $r[N, N] = s[N, N]$ and r normally generates N. One relator group theory (see Magnus, Karras, Solitar [MaKaSo76], Chapter 4, Section 4.4) implies that r is conjugate to $a^{\pm 5}$ in F, because a^5 also normally generates N. So we have $r = wa^{\pm 5}w^{-1}$, for some $w \in F$. The 4-term exact sequence (1.1) from Section 1.2 provides an embedding

$$N/[N, N] \xrightarrow{\partial} C_1(\Gamma(G, \mathbf{g})),$$

where the chain group on the right is a free $\mathbb{Z}G$-module with basis $\{\tilde{a}, \tilde{b}\}$. Let $\beta = \sum_{i=0}^{4} a^i \in \mathbb{Z}G$. We have

$$\partial(s[N, N]) = \alpha \, \partial(a^5[N, N]) = \alpha \beta \tilde{a},$$

and
$$\partial(r[N,N]) = \pm w\, \partial(a^5[N,N]) = \pm w\beta\tilde{a}.$$

Thus $r[N,N] = s[N,N]$ implies that $\pm w\beta = \alpha\beta$ in $\mathbb{Z}G$. Direct inspection shows that this last equation does not hold.

We can also give a more visual argument based on the isomorphism
$$N/[N,N] \xrightarrow{\partial} H_1(\Gamma(G,\mathbf{g})).$$

The Cayley graph $\Gamma(G,\mathbf{g})$ is a "tree of circles", each circle consisting of 5 edges labelled by a and at each vertex we have an incoming edge and an outgoing edge labelled by b. Shrinking each circle to a point gives an infinite regular tree, where each vertex has valency 10, with 5 incoming and 5 outgoing edges, all labelled by b. The 5-cycles provided by the circles in the Cayley graph form a $\mathbb{Z}$-basis for $H_1(\Gamma(G,\mathbf{g}))$. Note that if $v = v'a^j$, then $v(\beta\tilde{a}) = v'(\beta\tilde{a})$. This shows that the set of elements $v(\beta\tilde{a})$, where $v \in G$ and either $v = 1$ or v ends in $b^{\pm 1}$, is a $\mathbb{Z}$-basis for $H_1(\Gamma(G,\mathbf{g}))$. So $s[N,N] = r[N,N]$ implies that $\pm w' = \alpha$ in $\mathbb{Z}G$, where $w = w'a^j$, $j \in \mathbb{Z}$, $w' = 1$ or is an element of G ending in $b^{\pm 1}$. That last equation does not hold.

Dunwoody's construction uses one relator group theory and the fact that the group ring of a group with torsion contains non-trivial units. We do not know an example of a torsion free group where relation lifting fails.

Dunwoody provides us with an example of an almost presentation $P_0 = [a,b \mid s]$ of $G = \mathbb{Z}_5 * \mathbb{Z}$ where the relation module generator $s[N,N]$ can not be lifted. Thus by Theorem 1.3 $A(P_0) \neq C(P)$ for any presentation P of G. However, if $P = \langle a, b \mid a^5 \rangle$, then
$$C(P) \xrightarrow{\alpha} A(P_0),$$
multiplication by the unit α in every degree, is a chain complex isomorphism.

We end this section with an example of an almost presentation for a group G that is not a presentation, but can be lifted to one. Let F be the free group on a, b, c, d and let J be the normal closure of the elements $s_1 = a[a,b]$, $s_2 = b[b,c]$, $s_3 = c[c,d]$, $s_4 = d[d,a]$. Then F/J presents the Higman group $\hat{G}$ [Hig51], which is infinite, torsion-free, and does not admit finite quotients. Let
$$P_0 = [a,b,c,d \mid s_1, s_2, s_3, s_4].$$

Let G be the trivial group presented by

$$P = \langle a, b, c, d \mid a, b, c, d \rangle.$$

Let $N = F$. Then the $s_i[N, N]$ generate $N/[N, N]$, so P_0 is an almost presentation of the trivial group G but not a presentation, and $A(P_0) = C(P)$.

1.6 Solving the realization problem for some groups

The geometric realization problem has been solved for some classes of finite groups, finitely generated free groups, and the torus group. The finite group case is discussed in the Appendix to this chapter. In this section we will outline the solution for the infinite groups just mentioned.

An *algebraic 2-type* is a triple $(G, M, [\kappa])$ consisting of a group G, a $\mathbb{Z}G$-module M, and a cohomology class $[\kappa] \in H^3(G, M)$. Algebraic 2-types $(G, M, [\kappa])$ and $(G', M', [\kappa'])$ are *isomorphic* if there are maps

$$G \xrightarrow{\phi} G'$$

and

$$M \xrightarrow{\psi} M'_\phi,$$

where ϕ is a group isomorphism, and ψ is a $\mathbb{Z}G$-module isomorphism such that the cohomology classes $[\kappa]$ and $[\kappa']$ correspond under the homomorphisms

$$H^3(G', M') \xrightarrow{\phi^*} H^3(G, M'_\phi) \xleftarrow{\psi_*} H^3(G, M).$$

The $\mathbb{Z}G$-module M'_ϕ is obtained from the $\mathbb{Z}G'$-module M' by defining the G action as $gm' = \phi(g)m'$.

We can assign to an algebraic $(G, 2)$-complex A the algebraic 2-type

$$(G, H_2(A), [\kappa_A]),$$

where the class $[\kappa_A]$ is obtained in the following way: extend A to a projective resolution of the group G

$$\cdots \xrightarrow{\partial_4} P_3 \xrightarrow{\partial_3} A_2 \xrightarrow{\partial_2} A_1 \xrightarrow{\partial_1} A_0 \xrightarrow{\epsilon} \mathbb{Z} \longrightarrow 0.$$

Exactness implies that

$$P_3 \xrightarrow{\partial_3} H_2(A)$$

and we take $[\kappa_A] = [\partial_3] \in H^3(G, H_2(A))$. The following result is due to

MacLane and Whitehead [1] [MaWh50]. See also Sieradski's Chapter II in [I].

Theorem 1.8 *(MacLane, Whitehead) Algebraic $(G, 2)$-complexes are chain-homotopically equivalent if and only if their algebraic 2-types are isomorphic.*

We next state a direct consequence of Theorem 1.8.

Corollary 1.9 *Let A_1 and A_2 be two algebraic $(G, 2)$-complexes. Assume that $H^3(G, M) = 0$ for all $\mathbb{Z}G$-modules M. Then A_1 and A_2 are chain-homotopically equivalent if and only if the $\mathbb{Z}G$-modules $H_2(A_1)$ and $H_2(A_2)$ are isomorphic.*

A group G is called *aspherical* if it is the fundamental group of a finite aspherical 2-complex.

Theorem 1.10 *Let G be an aspherical group and assume that all finitely generated projective $\mathbb{Z}G$-modules are free. Let A_1 and A_2 be two finite algebraic $(G, 2)$-complexes and $\chi(A_1) = \chi(A_2)$. Then A_1 and A_2 are chain-homotopically equivalent.*

Proof: Since G is an aspherical group both $H_2(A_i)$ are finitely generated and projective by Schanuel's lemma. Hence they are both free. Now $\chi(A_1) = \chi(A_2)$ implies that they are both free of the same rank and hence are isomorphic. Corollary 1.9 implies the result. □

Corollary 1.11 *Suppose G is finitely generated and free, or $G = \mathbb{Z} \times \mathbb{Z}$. Let A_1 and A_2 be two finite algebraic $(G, 2)$-complexes and $\chi(A_1) = \chi(A_2)$. Then A_1 and A_2 are homotopically equivalent. Furthermore, any finite algebraic 2-complex is geometrically realizable.*

Proof: Finitely generated free groups and the torus group $\mathbb{Z} \times \mathbb{Z}$ satisfy the conditions of Theorem 1.10. See Cohn [Coh64] (see also Hog-Angeloni [I], [Ho-An90$_1$]) and Quillen [Qui76]. Thus the first part in the statement of the corollary follows from Theorem 1.10. We will prove geometric realization for the torus group. The arguments for the free group are similar. Consider the presentation $P = \langle a, b \mid [a, b] \rangle$. Note that $K(P)$ is aspherical: $\pi_2(K(P)) = 0$. Let A be a finite algebraic 2-complex. Then $H_2(A)$ is isomorphic to $\mathbb{Z}G^k$, for some k. Let K be the 2-complex obtained from $K(P)$ by wedging on k two-dimensional spheres. Then $\pi_2(K)$ is isomorphic to $\mathbb{Z}G^k$. It follows that $H_2(C(\tilde{K}))$ is isomorphic to $\mathbb{Z}G^k$. The result follows from Corollary 1.9. □

The next result is Corollary 4.3 in Harlander and Jensen [HaJe06].

Theorem 1.12 *Let G be a finitely presented aspherical group and let $N/[N,N]$ be the relation module associated with a finite generating set $\mathbf{g}$. Then there exists a finite presentation P of G so that the $\mathbb{Z}G$-module $\pi_2(K(P))$ is isomorphic to $N/[N,N]$.*

We describe the construction of the presentation P in the statement of the theorem. Choose aspherical presentations $Q = \langle \mathbf{x} \mid \mathbf{r} \rangle$. Choose a bijection $\phi \colon \mathbf{x} \to \mathbf{x}'$. A word w in $\mathbf{x}^{\pm 1}$ has a twin word w' in $\mathbf{x}'^{\pm 1}$ obtained from w by replacing an occurrence of $x^{\pm 1}$ by $\phi(x)^{\pm 1}$. We can form the twin presentation $Q' = \langle \mathbf{x}' \mid \mathbf{r}' \rangle$. Choose words u_g in $\mathbf{x}$ that represent $g \in \mathbf{g}$. Now define

$$P = \langle \mathbf{x} \cup \mathbf{x}' \mid \mathbf{r} \cup \mathbf{r}' \cup \{u_g u_g'^{-1} \mid g \in \mathbf{g}\}\rangle.$$

This presentation has the desired property. See [HaJe06] for details.

Corollary 1.13 *Let A be a finite algebraic $(G,2)$-complex, where G is a finitely presented aspherical group. If $H_2(A)$ is isomorphic to a relation module of G, then A is geometrically realizable.*

Proof: Suppose $H_2(A)$ is isomorphic to a relation module $N/[N,N]$. By Theorem 1.12 there exists a presentation P of G so that $\pi_2(K(P))$ is isomorphic to $N/[N,N]$. Thus $H_2(C(P))$ is isomorphic to $N/[N,N]$ and hence isomorphic to $H_2(A)$. By Corollary 1.9 A and $C(P)$ are homotopically equivalent. □

1.7 Exotic presentations of the trefoil group

Let G be the trefoil group presented by $P = \langle a,b \mid a^2 b^{-3}\rangle$. Since P is a 1-relator presentation for the torsion-free group G, the 2-complex $K(P)$ is aspherical (see Lyndon-Schupp [I], [LySc77]). Berridge and Dunwoody [BeDu79] showed that the relation module $N_i/[N_i,N_i]$ associated with the generating set $\mathbf{g}_i = \{a^{2i+1}, b^{3i+1}\}$, $i \in \mathbb{N}$, is stably free of rank 1 but not free. They showed furthermore that the set $\{N_i/[N_i,N_i]\}_{i\in\mathbb{N}}$ contains infinitely many distinct isomorphism types. It follows from Theorem 1.12 that there exist presentations P_i, $i \in \mathbb{N}$, so that the set $\{K(P_i)\}_{i\in\mathbb{N}}$ contains infinitely many distinct homotopy types of $(G,2)$-complexes with the same fundamental group and Euler characteristic. Dunwoody [I], [Du76] constructed a presentation P_1 with $\pi_2(K(P_1))$ isomorphic to $N_1/[N_1,N_1]$ without the use of Theorem 1.12. Lustig [I], [Lu93] used a construction similar to the one used in the proof of Theorem 1.12 to construct an infinite set of homotopically distinct finite 2-complexes with the same fundamental group and Euler characteristic.

In the remainder of this section we present some of the results just mentioned from the viewpoint of geometric realization. For simplicity of notation we do not distinguish between words in the free group on a, b and the elements these words present in G. The context will make these distinctions clear. Let $N/[N, N]$ be the relation module associated with the generating set $\{a, b\}$ of G. Since $K(P)$ is aspherical it follows that $N/[N, N]$ is free of rank 1 and $r[N, N]$ is a generator, where $r = a^2 b^{-3}$. Let $\alpha_i = 1 + a + \ldots + a^{2i}$ and $\beta_i = 1 + b + \ldots + b^{3i}$, $i \in \mathbb{N}$. These elements generate the left module $\mathbb{Z}G$. In order to see this observe that

$$(a-1)\alpha_i = a^{2i+1} - 1, \ (b-1)\beta_i = b^{3i+1} - 1. \tag{1.3}$$

Since

$$(a^{2i+1})^3 (b^{3i+1})^{-3} = a \text{ and } (a^{2i+1})^2 (b^{3i+1})^{-2} = b,$$

the group elements a^{2i+1} and b^{3i+1} generate G. It follows that $(a-1)\alpha_i$ and $(b-1)\beta_i$ generate the augmentation ideal IG. Since the augmentation map $\mathbb{Z}G \xrightarrow{\epsilon} \mathbb{Z}$ sends $3\alpha_i - 2\beta_i$ to 1, we see that $(a-1)\alpha_i$, $(b-1)\beta_i$, $3\alpha_i - 2\beta_i$, and hence α_i and β_i, generate $\mathbb{Z}G$. It follows that

$$\alpha_i \cdot r[N, N] = (r)(ara^{-1})\ldots(a^{2i} ra^{-2i})[N, N]$$

and

$$\beta_i \cdot r[N, N] = (r)(brb^{-1})\ldots(b^{3i} rb^{-3i})[N, N]$$

generate the relation module, where $r = a^2 b^{-3}$. Let

$$s_{1,i} = (r)(ara^{-1})\ldots(a^{2i} ra^{-2i}), \ s_{2,i} = (r)(brb^{-1})\ldots(b^{3i} rb^{-3i}).$$

Then $P_{0,i} = [a, b \,|\, s_{1,i}, s_{2,i}]$ is an almost presentation of G. The next result is a direct consequence of Berridge and Dunwoody [BeDu79].

Theorem 1.14 *(Dunwoody) The set $\{A(P_{0,i})\}_{i \in \mathbb{N}}$, contains infinitely many distinct chain-homotopy types of algebraic $(G, 2)$-complexes of Euler characteristic 1. Each algebraic $(G, 2)$-complex $A(P_{0,i})$ is geometrically realizable.*

Proof: Using the equations (1.3) one can show that $H_2(A(P_{0,i}))$ is isomorphic to the relation module $N_i/[N_i, N_i]$ associated with $\mathbf{g}_i$ (see Dunwoody [I], [Du76]). It follows that the set $\{A(P_{0,i})\}_{i \in \mathbb{N}}$ contains infinitely many distinct chain-homotopy types of algebraic $(G, 2)$-complexes. Corollary 1.13 shows that each $A(P_{0,i})$ is geometrically realizable. □

Dunwoody showed in [I], [Du76] that $P_1 = \langle a, b \mid s_{1,1}, s_{2,1} \rangle$ is a presentation of G, so $A(P_{0,1}) = C(P_1)$. It is probably true that a similar equation holds for all i, not just for $i = 1$, however we do not have a reference.

1.8 Exotic almost presentations for the Klein bottle group

By a *surface group* we mean the fundamental group of a compact surface. It follows from Section 1.6 that the geometric realization problem for surface groups has an affirmative answer in case the surface has boundary (and hence the group is free) or in case the surface is the torus. For other surface groups the geometric realization problem seems wide open. In this section we will display a result similar to Theorem 1.14 where the trefoil group is replaced with the Klein bottle group. Much of what was done in the previous section was based on the fact that the trefoil group admits an abundance of non-isomorphic relation modules associated with minimal generating sets. This however is not true for surface groups. Louder [Lou15] showed that any two finite generating systems of a non-trivial surface group are Nielsen equivalent. This implies the following result.

Theorem 1.15 *Relation modules associated with finite generating sets of surface groups are free.*

Key to the construction of exotic algebraic $(G, 2)$-complexes for the Klein bottle is the work of Artamonov [Art81] and Stafford [Sta85]. Given a Noetherian domain R and an automorphism $\sigma: R \to R$, one can define the skewed Laurent-polynomial ring $S = R[x, x^{-1}, \sigma]$, where we require $xr = \sigma(r)x$ for all $r \in R$. Stafford showed the following:

Theorem 1.16 *(Stafford) If $r_1, r_2 \in R$ satisfy the properties*

1 $S = S r_1 + S(x + r_2)$,
2 $\sigma(r_1)r_2 \notin R r_1$,

then the left ideal $K = \{s \in S \mid s r_1 \in S(x + r_2)\}$ is not generated by a single element.

Note that K is isomorphic to the kernel of the S-module epimorphism $S \oplus S \to S$, sending $(1, 0)$ to r_1 and $(0, 1)$ to $x + r_2$. Hence $K \oplus S \cong S \oplus S$. Since K is not generated by a single element, it is not free.

Let $P = \langle a, b \mid aba^{-1}b\rangle$. Then P presents the Klein bottle group G. The group ring $\mathbb{Z}G$ is a skewed Laurent-polynomial ring as $\mathbb{Z}G = \mathbb{Z}H[a, a^{-1}, \sigma]$, where $H = \langle b \rangle$ and $\sigma \colon \mathbb{Z}H \to \mathbb{Z}H$ is the ring isomorphism defined by $\sigma(b) = b^{-1}$.

Lemma 1.17 *Let $p(b)$ be a Laurent-polynomial in $\mathbb{Z}H$ and let $q(b) = p(b^{-1})$. Then $\alpha = a + q(b)$ and $\beta = p(b)$ generate $\mathbb{Z}G$ as a left $\mathbb{Z}G$-module.*

Proof: We have

$$(a - p(b))\alpha + p(b^{-1})\beta = (a - p(b))(a + p(b^{-1})) + p(b^{-1})p(b) =$$
$$a^2 + ap(b^{-1}) - p(b)a - p(b)p(b^{-1}) + p(b^{-1})p(b) = a^2.$$

The last equality used that fact that $ap(b^{-1}) = \sigma(p(b^{-1}))a = p(b)a$. □

If we let $p(b) = 1 + nb + nb^3$ and $q(b) = 1 + nb^{-1} + nb^{-3}$ play the roles of r_1 and r_2, respectively, then, using Lemma 1.17, we see that the conditions of Theorem 1.16 are satisfied, and it follows that the kernel K_n of the epimorphism

$$\mathbb{Z}G \oplus \mathbb{Z}G \xrightarrow{\psi} \mathbb{Z}G$$

sending $(1, 0)$ to $\alpha_n = a + 1 + nb^{-1} + nb^{-3}$ and $(0, 1)$ to $\beta_n = 1 + nb + nb^3$ is stably free of rank 1 but not free. Since P is a 1-relator presentation of the torsion-free group G, the 2-complex $K(P)$ is aspherical and the relation module $N/[N, N]$ associated with the generating set $\{a, b\}$ is free of rank 1, generated by $r[N, N]$, where $r = aba^{-1}b$. Let

$$s_{1,n} = (ara^{-1})(r)(b^{-1}r^n b)(b^{-3}r^n b^3), \quad s_{2,n} = (r)(br^n b^{-1})(b^3 r^n b^{-3}).$$

Note that

$$s_{1,n}[N, N] = \alpha_n \cdot r[N, N], \quad s_{2,n}[N, N] = \beta_n \cdot r[N, N],$$

so each

$$P_{0,n} = [a, b \mid s_{1,n}, s_{2,n}], \quad n \in \mathbb{N}$$

is an almost presentation of G.

Theorem 1.18 *The set $\{A(P_{0,n})\}_{n\in\mathbb{N}}$ contains infinitely many distinct chain-homotopy types of algebraic $(G, 2)$-complexes of Euler characteristic 1.*

Proof: Note that K_n is isomorphic to $H_2(A(P_{0,n}))$. Artamonov [Art81] shows that the set $\{K_n\}$ contains infinitely many distinct isomorphism types. His reasoning is as follows. First construct an infinite set of primes Q such that if $p < q$ and both p and q are in Q, then $q = 1$ modulo p. This is possible by

Dirichlet's Theorem. Let $K_{n,p} = K_n/pK_n$. Using Stafford's construction one can show that $K_{q,p}$ is not free for $p < q$, but $K_{q,q}$ is free. Thus if $q_1 < q_2$, then K_{q_1} and K_{q_2} are not isomorphic because K_{q_1,q_1} is free but K_{q_2,q_1} is not free. □

Unlike in the case of the trefoil group, we do not know if the algebraic $(G, 2)$-complexes are geometrically realizable.

1.9 Appendix: Geometric Realization for Algebraic 2-Complexes with Finite Fundamental Group

By F. Rudolf Beyl and Jens Harlander

In this appendix the following assumptions hold throughout: groups, 2-complexes, algebraic 2-complexes are assumed to be finite. These finiteness assumptions give our topic a decidedly number-theoretic flavor, through integral representation theory and algebraic K-theory arguments for $\mathbb{Z}G$ and closely related rings. Two recent books by F.E.A. Johnson ([Joh03], [Joh12]) make the background material quite accessible, and the reader is invited to consult these books for more information. In particular, a major theme of these books is the proof of Theorem 1.6 above.

1.9.1 Some general results

Given the group G, recall $\chi_{alg}(G, 2)$ and $\chi_{geom}(G, 2)$ from Section 1.2. If K is $(G, 2)$-complex, we denote by $[K]$ its homotopy class. If A is an algebraic $(G, 2)$-complex, we denote by $[A]$ its $\mathbb{Z}G$-chain homotopy class. Let

$$h_{geom}(m, G) = \#\{[K] \mid K \text{ is a } (G, 2)\text{-complex and } \chi(K) = m\}$$

$$h_{alg}(m, G) = \#\{[A] \mid A \text{ is an algebraic } (G, 2)\text{-complex and } \chi(A) = m\}.$$

We note that

$$h_{geom}(m, G) \leq h_{alg}(m, G).$$

This is because homotopically distinct $(G, 2)$-complexes K_1 and K_2 give rise to chain-homotopically distinct algebraic $(G, 2)$-complexes $C(\widetilde{K}_1)$ and $C(\widetilde{K}_2)$, as follows from Theorem 1.8.

Theorem 1.19 *(Browning [1], [Br79$_3$]) If $m > \chi_{geom}(G, 2)$, then $h_{geom}(m, G) = 1$. If $m > \chi_{alg}(G, 2)$, then $h_{alg}(m, G) = 1$.*

This theorem reduces the classification of both algebraic and geometric $(G, 2)$-complexes to the minimal possible Euler characteristic level. In particular we have:

Corollary 1.20 *Suppose $\chi_{geom}(G, 2) = \chi_{alg}(G, 2)$. If A is an algebraic $(G, 2)$-complex and $\chi(A) > \chi_{alg}(G, 2)$, then A is geometrically realizable.*

Proof: Let K be a $(G, 2)$-complex such that $\chi(K) = \chi_{geom}(G, 2)$. Let $K(n)$ be K with a bouquet of n 2-spheres wedged on. Since

$$\chi(A) > \chi_{alg}(G, 2) = \chi_{geom}(G, 2)$$

there exists an n so that $\chi(K(n)) = \chi(A)$. Let $C(\widetilde{K}(n))$ be the cellular chain complex of $\widetilde{K}(n)$, the universal covering of $K(n)$. Then $\chi(C(\widetilde{K}(n))) = \chi(A) > \chi_{alg}(G, 2)$, and hence these two algebraic $(G, 2)$-complexes are chain homotopically equivalent by Theorem 1.19. □

Corollary 1.21 *Let G be a finite group and $m = \chi_{geom}(G, 2)$. If we have $h_{geom}(m, G) > 1$ then $\chi_{geom}(G, 2) = \chi_{alg}(G, 2)$. In particular, no relation gap can occur at the minimal level.*

Proof: If $\chi_{geom}(G, 2) > \chi_{alg}(G, 2)$ then we would have more than one algebraic homotopy type above $\chi_{alg}(G, 2)$, contradicting Theorem 1.19. □

1.9.2 Cancellation Properties

Given augmented algebraic $(G, 2)$-complexes

$$A = (0 \longrightarrow M \longrightarrow A_2 \longrightarrow A_1 \longrightarrow A_0 \longrightarrow \mathbb{Z} \longrightarrow 0),$$

$$B = (0 \longrightarrow N \longrightarrow B_2 \longrightarrow B_1 \longrightarrow B_0 \longrightarrow \mathbb{Z} \longrightarrow 0).$$

By Swan's version of Schanuel's Lemma ([Swa60] Lemma 1.1),

$$M \oplus F_1 \cong N \oplus F_2, \tag{1.4}$$

where $F_1 = B_2 \oplus A_1 \oplus B_0$ and $F_2 = A_2 \oplus B_1 \oplus A_0$ are free modules of finite rank. In case A and B have the same Euler characteristic, F_1 and F_2 have the same rank, say k, thus

$$M \oplus \mathbb{Z}G^k \cong N \oplus \mathbb{Z}G^k. \tag{1.5}$$

If (1.4) holds for any free modules F_1 and F_2 of finite rank, M and N are called stably equivalent. If (1.4) holds for the trivial module $N = 0$, M is called stably free. In general, one must not infer $M \cong N$ from (1.5). However, if this conclusion holds for a particular group G, G is said to have *free cancellation*.

Recall that G is finite. The integral group ring $\mathbb{Z}G$ is said to satisfy *cancellation* if $\mathbb{Z}G \oplus P = \mathbb{Z}G \oplus Q$ implies $P = Q$ for any finitely generated projectives P and Q; and $\mathbb{Z}G$ is said to satisfy *stably free cancellation* if every finitely generated stably free $\mathbb{Z}G$-module is free. Clearly cancellation implies stably free cancellation.

A key tool in homotopy classification of algebraic and geometric $(G, 2)$-complexes is the result that $\mathbb{Z}G$ satisfies cancellation *unless* G admits a non-cyclic binary polyhedral group as a quotient (Swan (notes by E.G. Evans) [I], [Sw70], p.179). Thus one looks at the binary polyhedral groups as sources for interesting examples. To this end, Swan [Swa83] studied in great detail what types of cancellation may hold or fail for the group rings of the binary polyhedral groups.

When G is a finite 3-manifold group, invoking the *Swan modules* ([Joh03], §36) may obviate some cancellation arguments. These are the left modules $\langle N, k \rangle \subset \mathbb{Z}G$, and their $Hom(-, \mathbb{Z})$ duals, where $N = \sum_{g \in G} g$ and k is an integer relatively prime to the group order. They are projective.

Here are some direct consequences of Swan's results:

1 ([Swa83], Theorem I, p.66) Let G be a binary polyhedral group. Then $\mathbb{Z}G$ satisfies cancellation if and only if G is one of the seven groups $Q_8, Q_{12}, Q_{16}, Q_{20}, T_{24}, O_{48}, I_{120}$.
2 ([Swa83], Theorem 15.5, p.128) Cancellation holds for direct products $G = C_m \times L$, provided L has cancellation and m is odd and relatively prime to the order of L. By choosing L as one of the seven groups in the previous list and m as just specified, one has an infinite supply of 3-manifold groups $G = C_m \times L$ whose integral group rings satisfy cancellation.
3 If p is an odd prime, then all Swan modules for $G = Q_{4p}$ are free ([Swa83], Theorem 17.8, p.138).

1.9.3 Homotopy classification for finite 3-manifold groups

Here is a partial summary of results by Beyl, Latiolais, Waller [BeLaWa97], on the homotopy classification of the spines of compact closed 3-manifolds with finite fundamental group G.

Now Thurston's geometrization conjecture and the Poincaré conjecture have become theorems. Thus the finite groups that can act on the 3-sphere S^3 continuously and without fixed points, are the groups already known to allow fixed-point-free orthogonal actions. These are the cyclic groups C_n, the binary polyhedral groups, certain groups $D(2^k, 2\ell + 1)$ of order $2^k(\ell + 1)$ and $T(8, 3^k)$ of order $8 \cdot 3^k$, together with the direct products of any of the foregoing noncyclic groups and a cyclic group of relatively prime order.

The classification goes back to Threlfall and Seifert ([ThSe30], [ThSe32]). Milnor's list ([I], [Mi57] Theorem 2) with small free presentations for these groups is frequently quoted, also in [BeLaWa97].

Among the binary polyhedral groups are the quaternion groups of order $4n$ for $n \geq 2$ (also called binary dihedral or generalized quaternion groups):

$$Q_{4n} = \langle x, y \mid x^n = (xy)^2 = y^2 \rangle$$

Note that the relation $(xy)^2 = y^2$ is equivalent to $yxy^{-1} = x^{-1}$. Thus $x^n = yx^ny^{-1} = x^{-n}$.

Theorem 1.22 *Let K be a minimal $(G, 2)$-complex where G is a finite noncyclic fundamental group of a closed 3-manifold M. Suppose that $\mathbb{Z}G$ has stably free cancellation. Then K is homotopy equivalent to a spine of M. In particular $h_{geom}(m, G) = 1$ for all $m \geq \chi_{geom}(G, 2)$.*

The final statement follows from the affirmative answer to the Poincaré conjecture, which implies that 3-manifolds with isomorphic finite noncyclic fundamental group are homeomorphic. If K_1 and K_2 are minimal $(G, 2)$-complexes then, according to the first part of the theorem, both are spines of 3-manifolds M_1 and M_2 with the same fundamental group G. Hence $M_1 = M_2 = M$. So K_1 and K_2 are both spines of M and hence are homotopically equivalent.

In case G is finite cyclic we have $\chi_{geom}(G, 2) = \chi_{alg}(G, 2) = 1$ and there is exactly one homotopy type on each Euler-characteristic level. This follows from work of Olum [I], [Ol65]. Also see Chapter III by Latiolais in [I].

Theorem 1.23 *Let K be a minimal $(G, 2)$-complex where G is the finite fundamental group of a closed orientable 3-manifold M. Suppose $\pi_2(K)$ is singly generated and that all Swan modules for G are free. Then K is homotopy equivalent to a spine of M.*

Combine the theorems with Swan's results quoted in Subsection 1.9.2.

Corollary 1.24 *Let G be one of $Q_8, Q_{12}, Q_{16}, Q_{20}, T_{24}, O_{48}, I_{120}$ or the direct product of one of these groups with a cyclic group of relatively prime order. Then all minimal $(G, 2)$-complexes are homotopy equivalent to a spine of a 3-manifold.*

Corollary 1.25 *Let $G = Q_{4p}$ for an odd prime p. Then all minimal $(G, 2)$-complexes are homotopy equivalent to a spine of a 3-manifold.*

1.9.4 Nancy's Toy

Beyl and Waller [BeWa05], [BeWa08] construct algebraic 2-complexes over the binary dihedral groups $G = Q_{4n}$ for odd $n \geq 7$ that are "exotic", i.e., not homotopy equivalent to the familiar truncated geometric cell complexes resulting from the orthogonal actions alluded to in Subsection 1.9.3.

To this end the authors exhibit stably free nonfree $\mathbb{Z}G$-modules P of rank 1. Stabilizing isomorphisms $\Theta \colon \mathbb{Z}G \oplus \mathbb{Z}G \cong \mathbb{Z}G \oplus P$ and the boundary maps of the $(G, 2)$-complexes are given by explicit rectangular matrices. This makes the examples accessible to both theoretical and computational investigations, in particular, on the Realization problem and Wall's D(2)-problem. Here we give a taste of the work by summarizing the key properties of the smallest case $n = 7$, informally known as Nancy's Toy.

Treating the elements of $(\mathbb{Z}Q_{28})^\ell$ as row vectors $[z_1, \ldots, z_\ell]$, $\mathbb{Z}Q_{28}$-linear maps $f \colon (\mathbb{Z}Q_{28})^2 \to (\mathbb{Z}Q_{28})^\ell$ are described as $2 \times \ell$ matrices, acting on the row vectors $[z_1, z_2]$ by matrix multiplication on the right.

Define the left $\mathbb{Z}Q_{28}$-module P as the left ideal $\langle -3 + 4y, x + 1 \rangle$ of $\mathbb{Z}Q_{28}$. Then $\mathbb{Z}Q_{28}/P \cong \mathbb{Z}/25$ and $NP = \langle N \rangle$, where $N \in \mathbb{Z}Q_{28}$ is the sum all group elements. By Theorem 3.2 of [BeWa05], P is not $\mathbb{Z}Q_{28}$-free, essentially because P/NP is not singly generated as a $\mathbb{Z}Q_{28}$-module.

Let $\Phi\colon (\mathbb{Z}Q_{28})^2 \to (\mathbb{Z}Q_{28})^2$ be the 2×2 matrix with entries

$\Phi_{11} = x^7[1 + (1-x^7)y][1+x^{-5}] - [7 - 7x^7 - \Sigma^-][1 + (x^{-3}+x^3)x^5y]$,
$\Phi_{12} = [1 + (1-x^7)y][1 - (x^{-3}+x^3)y] + [7 - 7x^7][1+x^5]$,
$\Phi_{21} = x^7[7-7x^7][1+x^{-5}] - [19-20x^7][1-(1-x^7)y][1+(x^{-3}+x^3)x^5y] + 14\Sigma^-$,
$\Phi_{22} = [7 - 7x^7 - \Sigma^-][1 - (x^{-3}+x^3)y] + [19 - 20x^7][1-(1-x^7)y][1+x^5]$,

where $\Sigma^- = 1 - x + x^2 - x^3 + \cdots + x^{12} - x^{13}$.

By Theorem 3.3 of [BeWa05], the image of Φ is $\mathbb{Z}Q_{28} \oplus P \subset (\mathbb{Z}Q_{28})^2$. Let Θ be the restriction of Φ to its image, then $\Theta\colon \mathbb{Z}Q_{28} \oplus \mathbb{Z}Q_{28} \cong \mathbb{Z}Q_{28} \oplus P$ is a stabilizing isomorphism for P.

Theorem 4.5 and Proposition 4.4 of [BeWa05] combine to:

Theorem 1.26 *(Nancy's Toy) The following algebraic $(\mathbb{Z}Q_{28}, 2)$-complex is exotic:*

$$0 \longrightarrow P/NP \longrightarrow (\mathbb{Z}Q_{28})^2 \xrightarrow{\partial_2} (\mathbb{Z}Q_{28})^2 \xrightarrow{\partial_1} \mathbb{Z}Q_{28} \xrightarrow{\varepsilon} \mathbb{Z} \longrightarrow 0,$$

where ε is augmentation, $\partial_1 = \begin{bmatrix} x-1 \\ y-1 \end{bmatrix}$, and

$$\partial_2 = \Phi \cdot \begin{bmatrix} \sum_{k=0}^{6} x^k & -1-y \\ y - \sum_{k=0}^{5} x^k & 1+yx \end{bmatrix}.$$

Nancy Waller proposed that her toy is not geometrically realizable. She died in 2015 while working on a chapter for this book. This leaves the project "The geometric realization problem for algebraic 2-complexes" [BeWa13], joint with Rudolf Beyl, unfinished. Their program is to modify a given algebraic $(G, 2)$-complex A in a controlled manner within the same $\mathbb{Z}G$-chain homotopy type, as to lay the groundwork for testing possible counterexamples. The examples in [BeWa05], [BeWa08] have geometric 1-skeletons. Extending analogues of geometric notions such as skeletons, prolongations, and Q**-transformations to the second dimension, the authors [BeWa13] arrive at criteria for the realizability of algebraic homotopy types that appear more suited for machine computation.

1.9.5 Geometric realization for groups of period 4

We survey some results from Johnson [Joh03]. A group G is said to have the geometric realization property if every algebraic $(G, 2)$-complex can be geometrically realized. We paraphrase Theorem 62.1 in [Joh03], page 234:

Theorem 1.27 *Let G be a group that admits a free resolution of period 4. If G has free cancellation then the following statements are equivalent:*

1. *G has the geometric realization property;*
2. *G has a balanced presentation.*

Examples of groups that satisfy the hypothesis of this theorem and have balanced presentations are:

- cyclic groups C_n;
- dihedral groups D_{4n+2} of order 2 mod 4;
- direct products $D_{2n} \times C_m$, where n and m are odd and coprime;
- the binary tetrahedral, octahedral, and icosahedral groups T_{24}, O_{48}, I_{120};
- the exceptional quaternion (binary dihedral) groups $Q_8, Q_{12}, Q_{16}, Q_{20}$, with balanced presentations listed in Subsection 1.9.3;
- the direct product $Q_8 \times C_3$.

1.9.6 Geometric realization for the dihedral groups

Let D_{4n} be the dihedral group of order $4n$. Latiolais showed that both homotopy and simply homotopy types of $(D_{4n}, 2)$-complexes are determined by their Euler characteristic. See Theorem 4.3 and Corollary 4.5 in [I], [La91]. In [MaO'S13] Mannan and O'Shea also classified algebraic $(D_{4n}, 2)$-complexes and showed that these dihedral groups have the geometric realization property.

The dihedral group D_{4n} is presented by

$$P = \langle a, b \mid a^{2n} = b^2 = 1, aba = b \rangle.$$

Let K be the standard 2-complex built from P. Note that the presentation P is efficient and hence minimal: $\chi_{geom}(D_{4n}, 2) = \chi_{alg}(D_{4n}, 2) = 2$. Let $K(k)$, $k \geq 0$, be the 2-complex obtained from K by wedging on a bouquet of k 2-spheres.

Let $C(\widetilde{K}(k))$ be the cellular chain complex of the universal covering $\widetilde{K}(k)$ of $K(k)$.

Theorem 1.28 *(Mannan, O'Shea [MaO'S13]) Let A be an algebraic $(D_{4n}, 2)$-complex and assume $\chi(A) = 2 + k$, $k \geq 0$. Then A is chain homotopically equivalent to $C(\widetilde{K}(k))$. In particular the dihedral groups D_{4n} have the geometric realization property.*

Corollary 1.29 *Let L be a $(D_{4n}, 2)$-complex and assume $\chi(L) = 2 + k$, $k \geq 0$. Then L is homotopically equivalent to $K(k)$.*

2

Further Results concerning the Andrews-Curtis-Conjecture and its Generalizations

Cynthia Hog-Angeloni and Wolfgang Metzler

2.1 Introduction

This chapter carries on the insights of Chapters I and XII of [I]. We assume that the reader is familiar with 3-deformation types of finite 2-dimensional CW-complexes and their algebraic counterparts, namely Q^{**}-classes of finite presentations, as developed in Chapter I of [I].

Let us recall briefly some basic notions of [I], Chapter I, pages 45 and 46. The *Andrews-Curtis-Conjecture* (AC) claims that a contractible 2-dimensional CW-complex K^2 can be 3-deformed to a point. Its *generalization* (AC') is the conjecture that a simple homotopy equivalence between arbitrary finite complexes K_0, K_1 can always be replaced by a 3-deformation. And the *relative version* (rel AC') claims that a common subcomplex $L \subseteq K_0 \cup K_1$ can be kept fix during this 3-deformation. Obviously there is the hierarchy (rel AC') implies (AC') which in turn implies (AC).

The positive solution of the 3-dimensional Poincaré-Conjecture implies that a contractible 2-complex K^2, which is a 3-manifold spine, is Q^{**}-trivial, compare (62), page 45 in [I]. Indeed, a spine of a 3-ball can even be Q-trivialized without intermediate prolongations; this is a special case of a result on Heegaard-decompositions by Guangyuan Guo [Guo16] which we will expose in section 2.4. The general case of all of (AC), (AC') and (rel AC') is still open.

In section 2.2 we will summarize and extend results on projections of the free group of generators into test groups thereby excluding the possibility that finite test groups might be able to detect (AC')-counterexamples. Section 2.3 contains candidates for potential counterexamples to (rel AC') which "die" when the subcomplex is no longer assumed to be kept fixed during a 3-deformation. We also summarize techniques on 3-deformations and Q^{**}-transformations which generalize to the relative case.

In section 2.5, we draw the reader's attention to the survey article of Martin

Bridson [Bri15] which appeared after the authors agreed on the plan of the present book.

Most of the material of this chapter has been worked out by students of ours in their theses, see the bibliography. The content of this chapter is also relevant for the following one on TQFT-approaches.

2.2 Projections onto test groups

In order to study whether two presentations $\mathcal{P} = \langle a_1, \ldots, a_g \mid R_1, \ldots, R_h \rangle$ and $Q = \langle a_1, \ldots, a_g \mid S_1, \ldots, S_h \rangle$ are Q–equivalent, one may project the generators a_i of $F(a_i)$ onto a test group G^* and study the image-"vectors" $(\overline{R_1}, \ldots, \overline{R_h})$ with respect to Q–transformations. If the identity map on $F(a_i)$ induces an isomorphism between the groups defined by $\mathcal{P}$ and Q, then the normal closures $N(\overline{R_j})$ and $N(\overline{S_j})$ must coincide. This normal subgroup of G^* we call G. There is hope that it may be easier to distinguish Q-classes after projections rather than prior to them.

For this purpose we define the Q-*graph* $\Delta_h(G)$ with a vertex $(\overline{R_1}, \ldots, \overline{R_h})$ for each presentation and an unoriented edge for each elementary Q-move

$$(\overline{R_1}, \ldots, \overline{R_h}) \longrightarrow (\overline{S_1}, \ldots, \overline{S_h}).$$

If $(\overline{R_1}, \ldots, \overline{R_h})$ and $(\overline{S_1}, \ldots, \overline{S_h})$ are in distinct connected components of this graph, then $\mathcal{P}$ and Q aren't Q-equivalent.

It is of particular interest whether simple-homotopy type and 3-deformation type can be distinguished by these means. But one needs already subtle arguments in order to prove that certain test quotients must fail. One such result is already contained in [I], Chapter I, pages 370-371. There we show that simple-homotopy equivalent complexes can be 3-deformed (*normalized*) until their 1-skeleta coincide and corresponding presentations $\mathcal{P} = \langle a_1, \ldots, a_g | R_1, \ldots, R_h \rangle$, $Q = \langle a_1, \ldots, a_g | S_1, \ldots, S_h \rangle$ fulfill the condition

(1) $N = N(R_j) = N(S_j)$ and all $R_j S_j^{-1}$ are contained in the commutator subgroup $[N, N] = N^{(1)}$.

Moreover, by further finite Q-transformations of $\mathcal{P}$ one can achieve that

(2) all $R_j S_j^{-1}$ become higher commutators, i. e. elements of $N^{(n)}$ for $n \in \mathbb{N}$.

This excludes solvable groups as test quotients to disprove (AC'). As mentioned in [I], the (AC)-case of this result dates back to an unpublished paper [I], [Br76$_2$] of W. BROWNING.

Concerning (AC), in another unpublished paper [Bro76], Browning asked whether a *finite* group G can serve for this purpose. He obtained the criterion that such a group would have to fulfill:

(3)   *the minimal number $d(G)$ of normal generators of G is equal to the minimal number of generators required for G and for its abelianization $G/[G,G]$.*

Without knowledge of Browning's result, A. V. BOROVIK, A. LUBOTZKY and A. G. MYASNIKOV extended (3) in the (AC)-case to the following

Lemma 2.1 *Let G be a finite group and $h \geq max(d(G), 2)$. Then $(\overline{R_1}, \ldots, \overline{R_h})$ is Q-equivalent to $(\overline{S_1}, \ldots, \overline{S_h})$ exactly if the projections of these vectors to $G/[G,G]$ are Q-equivalent, in other words: The connected components of $\Delta_h(G)$ are the preimages of the connected components of $\Delta_h(G/[G,G])$.*

The clever inductive proof can be found in [BoLuMy05], see also [Hän05].

In this special case we have $G = G^*$. But the arguments of [BoLuMy05] can be extended to Q-transformations in the case of a proper inclusion $G^* \supset G$ step by step. Hence we can use their work also in the (AC')-case:

Theorem 2.2 *For normalized presentations it is impossible to detect counterexamples to (AC') by use of finite test quotients.*

Proof: Assume that we have presentations $\mathcal{P} = \langle a_1, \ldots, a_g \mid R_1, \ldots, R_h \rangle$ and $\mathcal{Q} = \langle a_1, \ldots, a_g \mid S_1, \ldots, S_h \rangle$ fulfilling (1). As G is the normal closure of the images of the defining relators, we have $h \geq d(G)$; and $h \geq 2$ may be assumed, because by Magnus, [I], [Ma30], $h = 1$ would imply $R_1 = wS_1^{\pm 1}w^{-1}$ in N which yields a Q-equivalence. Hence we may focus on the case $h \geq max(d(G), 2)$. Now (1) implies that $(\overline{R_1}, \ldots, \overline{R_h})$ and $(\overline{S_1}, \ldots, \overline{S_h})$ project to identical vertices in $\Delta_h(G/[G,G])$; hence they belong to the same component of the preimage by the extension of Lemma 2.1. □

Motivated by (1) and the extension of Lemma 2.1, where all commutators of relators finally are modded out, the authors want to draw attention to the following Theorem 2.3, which doesn't need a finite test quotient at all. It collects material from [I].

Recall that the abelianization of the normal subgroup $N \triangleleft F(a_1, \ldots, a_g)$ is the relation module $\mathbf{N}_\mathcal{P}$. We denote by $\overline{R_i}, \overline{S_j}$ the canonical images of R_i, S_j in $\mathbf{N}_\mathcal{P}$.

Theorem 2.3 *A homology equivalence, i.e. a geometric map* $f \colon K_\mathcal{P}^2 \to K_Q^2$ *which is the identity on the 1-skeleton inducing isomorphisms in π_1 and the second homology group, can be obtained by a Q-transformation* $(\overline{R_1}, \ldots, \overline{R_h}) \longrightarrow (\overline{S_1}, \ldots, \overline{S_h})$.

The $\overline{R_i}, \overline{S_j}$ are obtained by Reidemeister-Fox derivation and the Q-transformation by $\mathbb{Z}\pi_1$-matrix operations.

The **Proof** follows from [I], Chapter II, Theorem 3.8 and [I], Chapter III on the *bias invariant*, see also [Met00]. It uses the fact that the second homology groups are direct summands of the second integral chain groups and obtains an echelon form for the map induced by f. □

W. Browning works with chains of subgroups of a finite group with operators instead.

In [BoLuMy05] the authors raise the question for which (infinite) groups the statement of Lemma 2.1 is valid; they say that these groups fulfill the "generalized Andrews-Curtis Conjecture". Note that this terminology differs from ours. But such groups likewise can be excluded as testgroups for disproving (AC).

2.3 Fixed subcomplexes

On the level of group presentations, fixed subcomplexes during a 3-deformation give rise to fixed generators and to fixed defining relators. The latter and at least all the generators which appear in them are not to be changed during what we call *relative Q^{**}-transformations*. Let us start with our favorite example where $\mathcal{P}$ and Q both have two generators t and a and the common defining relator $T = [t(a), a^m]^m$ which has to be fixed throughout. $\mathcal{P}$ has the additional defining relator $R = t^j(a)[t(a), a^m]$ and Q has as additional defining relator $S = a$. Here and later we use the abbreviations $[x, y]$ for the commutator $xyx^{-1}y^{-1}$ and $x(y)$ for the conjugate xyx^{-1}. m and j are natural numbers which will be subject to restrictions. In short we denote this situation as

(4) $\mathcal{P}: \dfrac{t^j(a)[t(a), a^m]}{[t(a), a^m]^m}$ $\quad Q: \dfrac{a}{[t(a), a^m]^m}$

Note that by T, the right part of R - and thus a, too - has order dividing m, now R implies that a becomes in fact trivial in $F(t, a)/N$. Thus, $\mathcal{P}$ and Q both define the group $\mathbb{Z}$ with essential generator t. In fact, there exists a Q-transformation from $\mathcal{P}$ to Q starting with modifications of T by R. But we can show:

Theorem 2.4 *If $m \geq 2$ and $j \neq 0, \pm 1$ then there is no Q-transformation from $\mathcal{P}$ to Q rel t, a, T.*

Proof: The relator subgroup $N \leq F(t, a)$ is generated by the t-conjugates of a. If we pass from N to Reidemeister's quotient $N/[N, [N, N]]$, see [I], [Re36], these t-conjugates commute with their commutators and R becomes equivalent to $t^j(a)[t(a), a]^m$. By a finite sequence of elementary Q-transformations (inversion, conjugation with $t^{\pm 1}$ or $a^{\pm 1}$, multiplication with T), R may be transformed to become equivalent to at most

(5) $t^k(a)^{\pm 1} \prod [t^k(a), t^i(a)]^{z(k,i)} \prod_{r, s \neq k} [t^r(a), t^s(a)]^{z(r,s)}$

with integral exponents $z(k, i)$, $z(r, s)$. But as $j \notin \{-1, 0, 1\}$ and $m \geq 2$, the changes from R to (5) imply that the factor $[t(a), a^m]$ gives rise to exactly one $z(r, s)$, $r, s \neq k$, which is not congruent 0 mod m^2. (Here we use that products mod $[N, [N, N]]$ like (5) can be considered to be in normal form where each ordered pair (k, i) and each (r, s) occur exactly once, see [I], [Re36] and [Küh00a]). Hence, R cannot be Q-transformed into S rel t, a, T. □

The preceding argument mod $[N, [N, N]]$ already breaks down, when $\mathcal{P}$ and Q are prolonged by a single additional generator b to

$\mathcal{P}': \dfrac{t^j(a)[t(a), a^m], b}{[t(a), a^m]^m}$ $\quad Q': \dfrac{a, b}{[t(a), a^m]^m}$

Mixing Q-transformations with equivalences ($\sim$) mod $[N, [N, N]]$ and abbreviating $[t^{-j+1}(x), t^{-j}(R)]^m$ for $x \in N$ by $K[x, R]$, we obtain the transitions

(6) $$\begin{aligned}(R,b) &\to (R \cdot K[b,R],b) \to (R \cdot K[b,R], R \cdot b \cdot K[b,R]) \tilde\to \\ & (R \cdot K^{-1}[R,R], R \cdot b \cdot K[b,R]) \to \\ & (R \cdot K^{-1}[R,R], b \cdot K[b,R] \cdot K[R,R]) \tilde\to (R \cdot K^{-1}[R,R], b).\end{aligned}$$

But $K^{-1}[R,R]$ is equivalent to $[t(a),a]^{-m}$. Hence we ended up with a pair equivalent to $(t^j(a), b)$ which is quite as good as (S, b). The generators t, a, b and T have not been touched during this Q-transformation and we have proved:

Theorem 2.5 *Up to modification mod* $[N, [N, N]]$, $\mathcal{P}'$ *and* Q' *can be Q-transformed into each other rel* t, a, b *and* T.

The calculations which we have carried out for our favorite example are consequences of a stronger result of ALEXANDER KÜHN, see [Küh00a]. He derives a relative version of section 2.2, (2), namely

(7) for $\mathcal{P} = \langle a_1, \ldots, a_g \mid R_1, \ldots, R_h, T_1, \ldots, T_\ell \rangle$
and $Q = \langle a_1, \ldots, a_g \mid S_1, \ldots, S_h, T_1, \ldots, T_\ell \rangle$ with $R_j S_j^{-1} \in N^{(1)}$
there exists a Q-transformation of $\mathcal{P}$ fixing all generators and $T_1, \ldots, T_\ell$, such that afterwards the transformed $R_1', \ldots, R_h'$ fulfill $R_j' S_j^{-1} \in N^{(n)}$ for $n \in \mathbb{N}$ provided $h \geq 2$.

Thus the different outcomes of Theorem 2.4 where $h = 1$ and Theorem 2.5 with $h = 2$ are not accidental. Note that these Theorems work with the start of the lower central series whereas (2) and (7) concern higher commutators.

We have started this section with a conjectured counterexample to rel (AC'); here are some more:

(8) $\mathcal{P}: \dfrac{a = [t^2(a), t(a)]}{[[t^3(a), t^2(a)], [t^2(a), t(a)]]}$ $Q: \dfrac{a}{[[t^3(a), t^2(a)], [t^2(a), t(a)]]}$

and a whole family:

(9) $\mathcal{P}: \dfrac{a}{a^{-1}b[\varphi(a), \psi(b)] = 1}$ $Q: \dfrac{b}{a^{-1}b[\varphi(a), \psi(b)] = 1}$

where the defining relators this time are denoted as equations, φ and ψ are additional variables, and $\varphi(a) = \varphi a \varphi^{-1}$, $\psi(b) = \psi a \psi^{-1}$ are abbreviations as ahead of (4).

For these examples it likewise holds that it is not possible to Q-transform the variable relator R into S by a consequence of the fixed one T (exercise: use normal forms). But we also have to take into account 2-dimensional expansions (=prolongations) introducing additional generators $a_1, \ldots, a_n$. As it is possible to lump the multiplication of consequences of T to the end or the beginning of a relative Q-transformation of the expanded presentations we would get a Q-transformation

(10) $\quad (Rf, a_1 f_1, \ldots, a_n f_n) \to (S, a_1, \ldots, a_n)$

with certain products f and f_i of conjugates of T. In particular we would get an isomorphism for the groups resp. complexes where T is suppressed resp. punctured.

One of our strategies to disprove (rel AC') is to show that for no choice of the modifications f and f_i can such an isomorphism exist. Note that the conjugates which are used in forming these modifications contain $a_1, \ldots a_n$. Nevertheless in the examples (4) and (8) we have been successful for certain families of modifications, see [Ho-AnMe04] and [Ho-AnMe06] as they give rise to t-layered coverings. Also the example from [I], page 379 (19) and the one given in (9) above deserve further attention.

As to test quotients in the relative case: By (7) one should refrain from solvable ones. But we haven't studied finite test groups and other ideas of [BoLuMy05] in this context so far.

In [Küh00b] there are more techniques that are established resp. should be treated in the relative case. A. Kühn proves that the n-dimensional version of P. WRIGHT's lemma (see [I], page 23) also holds if a subcomplex is kept fix:

(11) If K_0 is transformed into K_1 by expanding and collapsing $(n+1)$-cells, and if L is a subcomplex of $K_0 \cup K_1$, then the expansions and collapses can be chosen transiently and fixing L throughout the deformation.

He also suggests a relative version of the Theorem on semisplit Q^{**}-transformations in [I], page 377, Theorem 3.4, see Chapter 7, Section 6, remark after the proof.

2.4 (AC) for spines of 3-manifolds

Every compact, connected, m-dimensional p.l. manifold M collapses to a spine of dimension $\leq m - 1$, see [I], page 31; in case M is closed, remove an m-ball first. This fact can be used to establish the existence of a Heegaard-diagram $(V; k_1, \ldots k_h)$ of a 3-dimensional orientable manifold M, with Heegaard-surface $F = \partial V$ and a complete system $k_1, \ldots k_n$ of pairwise disjoint simple closed curves on F such that $F \setminus \bigcup_{i=1}^{n} k_i$ is a sphere with holes.

Equivalence is defined using two elementary operations (and their inverses):
Equivalence of Heegaard-decompositions (M, F) and (M, F') of a 3-manifold M means that there exists a homeomorphism $h : M \to M$ isotopic to the identity of M which induces an orientation preserving homeomorphism of F onto F'.

Equivalence of Heegaard-diagrams $(V; k_1, \ldots k_n)$ and $(V; k'_1, \ldots k'_n)$ is generated by the elementary operation of sliding some k_i over some k_j along some connecting curve whose interior meets no k_ℓ at all, while all others k_ℓ are fixed, to obtain k'_i; or $(V; k'_1, \ldots k'_n)$ is obtained from $(V; k_1, \ldots k_n)$ by applying some homeomorphism to V which maps $k_1, \ldots k_n$ to $k'_1, \ldots k'_n$.

Recently, GUANGYUAN GUO, (see [Guo16]), has shown:

Theorem 2.6 *Assume* $\mathcal{P} = \langle a_1, \ldots, a_n \mid R_1, \ldots, R_n \rangle$ *is a presentation of the trivial group read off a spine* K^2 *of a 3-dimensional closed manifold* M^3. *Then* $\mathcal{P}$ *is Q-equivalent to the trivial presentation* $\mathcal{P} = \langle a_1, \ldots, a_n \mid a_1, \ldots, a_n \rangle$.

The proof comes out of two big Theorems: Firstly, Perelman's Theorem (see [Per02], [Per03b], [Per03a]) which implies that M^3 must be in fact the 3-sphere. Secondly, WALDHAUSEN's Theorem, (see [Wald68]) which implies that any HEEGAARD-decomposition (S^3, F) of the 3-dimensional sphere with Heegaard-surface F is equivalent to the trivial one of same genus.

Guo proves more generally, that for closed 3-manifolds, equivalence classes of Heegaard-decompositions of genus n are in one-to-one correspondence with equivalence classes of Heegaard-diagrams. Further, the curves $k_1, \ldots k_n$ give rise to elements of $\pi_1(V)$ (up to conjugation) such that for equivalent Heegaard-diagrams the corresponding sets of curves are Q-equivalent. Thus, in the case of a balanced presentation of the trivial group, the given presentation $\mathcal{P} = \langle a_1, \ldots, a_n \mid R_1, \ldots, R_n \rangle$ of the 3-manifold-spine is Q-trivial.

In the context of Guo's work we also recommend the thesis [Li00] of ZHONG-MOU LI.

2.5 Considerations on length

In [Bri15], Martin Bridson has taken up the considerations on length for (AC) in the context of *Dehn functions*, see [I], Chapters VI and XII. He obtains large lower bounds on the minimum numbers of (certain) Q-moves which are necessary in order to trivialize the members of a family $\mathcal{P}_n$ of balanced presentations of $\pi = 1$. The sum of the relators of $\mathcal{P}_n$ is at most $24(n + 1)$ but the number of moves of the corresponding Dehn function "grows more quickly than any tower of exponentials; in particular it quickly exceeds the number of electrons in the universe" (citation), although these presentations are Q-trivial.

The paper contains hints to topological and geometric consequences and promises to discuss details of these in a sequel; see also [Lis17] and [LiNa17].

2.6 On the future of the Andrews-Curtis problem

a) The hierarchy (rel AC') $\Rightarrow$ (AC') $\Rightarrow$ (AC) mentioned in section 2.1 and the examples in section 2.3 of this article suggest the prognosis that the first one is most likely to be disproved. There are also invariants of standard algebraic topology which might be useful for it.

b) As Bridson focuses on algorithmic properties, these might precede a solution of (AC) as happened in the case of the 3-sphere and the Poincaré-Conjecture in dimension 3, see SERGEI MATVEEV's book [Mat03].

c) Algebraic considerations on the Andrews-Curtis problem lead to isomorphism questions of bases and lifts (see [I], Chapter XII, Section 2.2), whereas the *relation gap problem* (see Chapters 6 and 7) "just" deals with the question of whether a small generating system of a relation module lifts to the relator subgroup of a presentation.

    Thus, after all, we consider the Andrews-Curtis-problem to be the harder one.

3
Aspects of TQFT and Computational Algebra

Holger Kaden and Simon King

3.1 Introduction

We start by introducing the basic concept and motivation of TQFT. The focus of the first sections is on the Quinn invariant for 2-complexes. To compute the Quinn invariant, the 2-complex is sliced into graphs, to which rooted trees are assigned. Their data are associated to semisimple tensor categories, based on Quantum groups. The combinatorial data give rise to a state sum polynomial, and the theory of quantum groups provides a way to evaluate the polynomials so that the result is invariant under local transformations. An overview of quantum invariants with their underlying algebraic setting and topological theorems can be found in [BlTu06b].

A potential approach to study simple homotopy equivalence is via TQFTs for so-called s-moves; this is discussed in Section 3.4.

A state sum invariant on compact 3-manifolds M^3 is constructed in [TuVi92]. This so-called *Turaev–Viro invariant* (TV invariant) has originally been expressed in terms of colourings of triangulations of 3-manifolds, using the $6j$-symbols obtained from a quantum group. But it can equivalently be expressed in terms of special spines of 3-manifolds, as in Section 3.5 below. One can show that a special colouring recovers classical invariants like $H_1(\partial X; \mathbb{Z}_2)$ and $H_2(X, \partial X; \mathbb{Z}_2)$.

A different construction of an invariant due to Reshetikhin and Turaev is based on the Kirby calculus: Closed orientable 3-manifolds can be obtained by Dehn surgery and can thus be represented by links with surgery data associated to the link components, and two representations of the same manifolds are related by Kirby moves. The *Reshetikhin–Turaev invariant* is closely related to the TV invariant, see for example [Tur94, Theorem 4.1.1] or [Vir10].

Computational algebra comes into play if one avoids evaluation of the polynomials: Using Gröbner bases, one obtains a so-called "ideal invariant" by determining a normal form of the state sum subject to polynomial relations corresponding to the local transformations. Experiments show that ideal Turaev-Viro invariants are stronger than quantum Turaev-Viro invariants.

Bobtcheva and Quinn (2005) show that Andrews-Curtis invariants obtained by tensoring a *modular* TQFT invariants (including the above-mentioned quantum invariants) cannot detect counterexamples to the Andrews–Curtis conjecture. However, it is known that ideal invariants of 4-thickenings obtained from a generalization of Turaev–Viro invariants are not modular. Hence, it is still possible that they can detect Andrews-Curtis counterexamples.

3.2 The Concept of Topological Quantum Field theory (TQFT)

The study of TQFT can be motivated by the creation of new invariants for solving problems which occur for classical group valued invariants on 2-complexes respectively differentiable 4-manifolds. Attaching S^2 respectively $S^2 \times S^2$ in the according dimensions yields stabilization phenomena involving undesired cancellation effects on those invariants. Thus they become useless for detecting counterexamples to (AC) or to (AC'), see Chapter 2, Section 1, or to the diffeomorphism problem (see [Qui92], [Mül00]). Since TQFT invariants live in a ring R, these effects can be suppressed. We give a brief introduction to the concept of TQFT. The left part of Figure 3.1.a) illustrates the topological aspect of TQFT: One considers certain classes of topological objects X, so-called *spacetime* or *bordism* with boundary objects Y_i. Examples include pairs of objects ($d+1$-dimensional manifolds, d-dimensional manifolds) or (2-complexes, graphs). A part of the boundary (here: $Y_1 \cup Y_2$) is called *incoming*, another part (here: Y_3) is named *outgoing*, and the spacetime $X: Y_1 \cup Y_2 \longrightarrow Y_3$ is thought of as a transition from the incoming to the outgoing boundary. As in Morse theory, X is sliced into *elementary bordisms*, say, X_i.

Definition 3.1 *A TQFT Z over a unital, commutative ring R associates R-modules (state modules) to boundary objects, and R-module homomorphisms to the elementary bordisms with the following properties:*

1. *The state module $Z(Y_1 \cup Y_2)$ associated with the disjoint union of boundary components is the tensor product $Z(Y_1) \otimes_R Z(Y_2)$, and the analogues are the disjoint union of spacetimes.*
2. *The gluing of bordisms as depicted in Figure 3.1.b) induces a composition of the according homomorphism; if $X_1: Y_1 \longrightarrow Y_2$, $X_2: Y_2 \longrightarrow Y_3$ and $X = X_1 \cup_{Y_2} X_2: Y_1 \longrightarrow Y_3$, then $Z(X) = Z(X_2) \circ Z(X_1)$.*

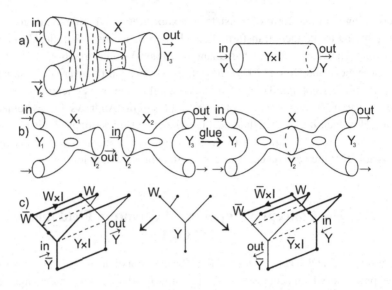

Figure 3.1 Properties of TQFT and induced module structures

3. The product $Y \times I: Y \longrightarrow Y$ induces the identity homomorphism on $Z(Y)$, depicted on the right of Figure 3.1.a).

The abstract concept is adopted from the *cobordism category of oriented, differentiable manifolds* $\mathcal{SDIFF}$. A pair of categories $(\mathcal{M}, \partial\mathcal{M})$ is defined, to include more general spaces; $\mathcal{M}$ now becomes spacetimes and $\partial\mathcal{M}$ are boundaries with functors $\partial: \mathcal{M} \longrightarrow \partial\mathcal{M}$ and $\times I: \partial\mathcal{M} \longrightarrow \mathcal{M}$. This is used to define a *collar* on the boundary:

- $X \cup_{\partial X} \partial X \times I \cong X$ for a spacetime X.
- $Y \times I \cup_Y Y \times I \cong Y \times I$ for a boundary Y.

In both categories there is an involution *bar*: $Y \longrightarrow \bar{Y}$, which reverses the orientation of Y. In particular, for the spacetime $X: Y_1 \longrightarrow Y_2$ the incoming boundary Y_1 is labelled by $\overline{Y_1}$. We get a *bordism category Bord($\mathcal{M}$)*, if we determine, when two spacetimes X, X' with the same incoming and outgoing boundary objects are defined to be equivalent. The definition of the equivalence relation itself depends on the category, where the TQFT will be constructed. Therefore an alternative definition of TQFT is possible:

Definition 3.2 *A TQFT is a functor* $Z: \text{Bord}(\mathcal{M}) \to \text{Mod}(R)$.

The functor Z assigns to each morphism class $[X]$ in Bord($\mathcal{M}$) the same homomorphism $Z(X)$, so Z defines an invariant. Setting $Z(\emptyset) = R$ and thinking of

Aspects of TQFT and Computational Algebra 39

2-complexes or closed manifolds as bordisms $\emptyset \to \emptyset$, Z assigns to a bordism a map $Z(\emptyset) \to Z(\emptyset)$; therefore it is an element of $Z(\emptyset) = R$ itself. A *modular* TQFT permits intersections of the incoming and the outgoing boundary into so called *corners* (these are objects in a category $\partial^2 \mathcal{M}$) and also a subdivision of the incoming boundary by corners; a corner ($n-1$-cell) subdivides a boundary (n-cell). However it is not allowed to subdivide the outgoing boundary by corners! For example, corners in Figure 3.1.c) are boundary points of and points inside a graph, where we have picked a boundary point of the boundary object, Y, denoted by W. We focus on gluing boundaries along corners. Let Y_1 and Y_2 be boundary objects and U the common subset of corners, which we glue together. Then the gluing is defined by an isomorphism:

$$Z(Y_1 \cup_U Y_2) \cong Z(Y_1) \otimes_{Z(U \times I)} Z(Y_2)$$

or

$$Z\left(\bigcup_U Y\right) \cong \bigotimes_{Z(U \times I)} Z(Y)$$

for a common boundary Y. Apply the functor $\times I : \partial^2 \mathcal{M} \to \partial \mathcal{M}$ for a corner $W \in \partial^2 \mathcal{M}$, where the boundary object $W \times I$ can be endowed with a ring structure on $Z(W \times I)$, see below.

Definition 3.3 *A modular TQFT is a TQFT with corners, that are objects $W \in \partial^2 \mathcal{M}$, so that the above isomorphism exists. It is required that the rings $Z(W \times I)$ and $Z(\partial Y \times I)$ possess a unit which acts by the identity on the state module $Z(Y)$.*

The modularity condition leads to a right module structure on $Z(Y_1)$, a left module structure on $Z(Y_2)$, which agree on $Z(Y)$. The isomorphism induces the quotient with respect to these structures. We explain the different module structures on $Z(Y)$, using Figure 3.1.c). The picture in the middle shows a boundary object (graph) Y with corners as its boundary points. Module structures are generated by bordisms and their induced homomorphisms. The left picture presents the bordism $Y \times I$ with incoming boundary $\bar{Y}$ and outgoing boundary Y. Pick up a corner $\overline{W} \in \bar{Y}$, interpret the attached segment $W \times I$ (including the direction of the arrow) on the corner $\overline{W}$ in $\bar{Y}$ as a collar of $\overline{W}$ (for all corners). We obtain a graph $\bar{Y} \equiv \bar{Y} \cup W \times I$ with the modular bordism $\bar{Y} \cup Y : \bar{Y} \to Y$. By considering for $W \in \partial Y$ the segments $W \times I \in \partial Y \times I$ separately, there is a bordism $\bar{Y} \cup (\partial Y \times I) \cup Y$, which induces a homomorphism: $Z(Y) \otimes Z(\partial Y \times I) \to Z(Y)$; hence $Z(Y)$ is equipped with a right module structure over $Z(\partial Y \times I)$, created from each structure $Z(W \times I)$ of the corners $\overline{W}$ in $\bar{Y}$. Analogous considerations for $\bar{Y} \times I$ result in a modular bordism $Y \cup \bar{Y} : Y \to \bar{Y}$

and a bordism $\partial \bar{Y} \times I \cup Y \cup \bar{Y}$, so $Z(Y)$ being equipped with a left module structure over $Z(\partial \bar{Y} \times I)$ (see the right picture of Figure 3.1.c), where $Y = \bar{\bar{Y}}$, and use $Y \equiv \overline{W} \times I \cup Y$).

Remark 3.4 *The above procedure explains the data in the gluing isomorphism, but it would involve further arguments on collaring as above, performed with respect to the bordism $Y_1 \cup_W Y_2 \to Y_1 \cup Y_2$ in order to show that there is a well defined homomorphism: $Z(Y_1) \otimes_{Z(W \times I)} Z(Y_2) \to Z(Y_1 \cup_W Y_2)$. See [Qui95].*

We get on $Z(W \times I)$ the structure of a (special) ambialgebra (Frobenius algebra). Using the bordism $W \times I^2$, the four boundary components $W \times I$ can be oriented and glued together to create the structure maps defining an *ambialgebra*. For example, the ring structure of $Z(W \times I)$ is obtained by the choice of two copies of $W \times I$ as incoming, and two glued copies of $W \times I$ as outgoing boundary. This induces the homomorphism $Z(W \times I) \otimes Z(W \times I) \longrightarrow Z(W \times I)$ in the sense of the remark. The ambialgebra relations can be confirmed by showing the equivalence of the underlying bordisms, composed by the bordisms of the structure maps, see [Qui95], [Mül00]. For an overview on the axiomatic world of TQFT, see [BlTu06a].

The concept of a TQFT is related to attempts in mathematical physics to unify general relativity and quantum mechanics to a common theory of quantum gravity: Since mass results in curvature in 4-dimensional spacetime, it is modeled by a 4-dimensional manifold M^4. Our 3-dimensional space at a given time corresponds to a slice of M^4, and the transition from one slice to another, *i.e.*, the progress of time, is a bordism. Since the formalism of quantum mechanics can be expressed in terms of operator algebras, one wants to express bordisms algebraically, as in a TQFT. We refer to [Poe13] for more details on that topic.

3.3 TQFT on 2-complexes

In [Qui92] Frank Quinn explores a new type of Andrews-Curtis invariant by constructing a TQFT on sliced 2-complexes. We consider the 2-complex sliced into a 1–*parameter family of graphs*. Studying its singularities, we get a list of local moves between the graphs, e.g. a circle to a point and vice versa. Note that Quinn [Qui92] prefers to use bordisms to the empty set, where we use the one-point set. Also by classifying the singularities for the transition of two 1–parameter families (the 2-complex and its deformed one), a list of topological relations among the local moves can be worked out, e.g. the (sliced) rectangle and its deformed rectangle with a bubble. In other words, the relations define two sequences of local moves to be equivalent! For the bordism 2-complex, associated R-modules and R-homomorphisms have been constructed. This determines a well-defined TQFT on the Andrews-Curtis transformation class of 2-complexes (see [I]), if equivalent sequences of local moves get equal homomorphisms. So far, these Andrews-Curtis invariants via TQFT for 2-complexes (due to Frank Quinn) still did not succeed in distinguishing between 3–deformations and simple homotopy equivalences. We apply results from Chapter 2, which shed some light on the Quinn invariant. For details see the standard references [Qui95], [Mül00]. In particular, one should consider nontrivial π_1 and as a next step fixed subcomplexes.

3.3.1 The Quinn model of a 2-complex

Let $P = \langle a, b \mid R, S \rangle$ be a presentation of a standard 2-complex K^2. Consider the generator circles as bases of cylinders connected by a rectangle; attach the 2-cell R according to the appearance of the generators from bottom to top, and close the curve by a connecting arc on the rectangle. Note that this can be achieved by 2-expansions on the generators and further 3-deformations. The 2-complex is in general position, the dots mark the centers of local vertex models, see Figure 3.2 and also Figure 3.4 for a sliced vertex model. The attached 2-

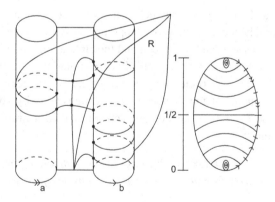

Figure 3.2 Attachment of a 2-cell $R = b^3 a^{-2} b^{-1}$ in the Quinn model

cells induce a height function on the (whole) 2-complex, starting with level 0 and ending with level 1. That corresponds to a decomposition of the 2-complex into a sequence of graphs (slicing or *sliced 2-complex*): It starts at the entry of the 2-cell (from the empty set) to a circle, which splits into an arc, one end on the connected rectangle, the other one on the circulating curve. This arc is called the *relator arc*; it slides along the 2-cell boundary and disappears at the exit of the 2-cell. The abstract 2-complex embeds into the 4-dimensional space, each slice with level t of the heightfunction embeds into the 3-dimensional subspace, see Figure 3.3.a). We get two circles connected by a segment (from

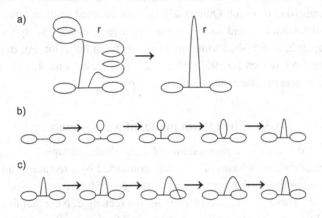

Figure 3.3 Slices in the Quinn model

the rectangle) and the relator arc, which is an unknotted curve in $\mathbb{R}^3$, free from selfintersections; hence it can be viewed topologically as a simple arc. We have added the 3 phases start/end of a 2-cell and the circulation around a generator in a relation, see Figure 3.3.b) respectively c). Caution: the slicing fulfills the Quinn list of local moves (potentially by adding certain flanges); but it has to be modified depending on the change of the attaching curves by a deformation. We illustrate the T_3 move as an example compare [I]: A theorem of Matveev [I], [Ma87$_1$] states:

Theorem 3.5 *Two special polyhedra P, Q can be 3-deformed into each other if and only if this transformation can be achieved by a sequence of moves $T_i^{\pm 1}$, $i = 1, 2, 3$.*

The left part of Figure 3.4 shows the transitions of local models by the T_i moves. Especially the T_3 move has two variations; starting with the slicing of its underlying local model, depicted in Figure 3.4.a). We indicate the slicing by the level of the slices on the bottom component. Figure 3.4.b) shows the slicing by changing the attaching curve of the bottom component by a turn to the right and Figure 3.4.c) by a turn to the left. The slicing in Figure 3.4.b) is, in an informal meaning, compatible with that of Figure 3.4.a), however not the slicing in Figure 3.4.c). This non-compatibility is an obstruction to ensure the Andrews-Curtis invariance for potential invariants. See [Kad10] for a detailed treatment.

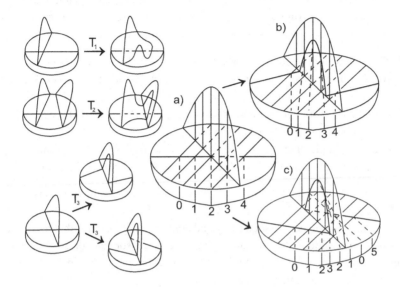

Figure 3.4 Vertex model with good and bad T_3 move

3.3.2 The algebraic concept

The start/end of a relation depicted in Figure 3.3.b) uses local moves in the Quinn list, however these processes are the same for all 2-complexes. The tool to distinguish 2-complexes is:

Definition 3.6 *The algebraic description of reading a generator in a relation is called the* circulator. *In more detail, it describes the relator arc slided around the generator as depicted in Figure 3.3.c).*

One should avoid using identities different from those generated by the list of topological relations. So we are looking for an algebraic structure expressed by manipulating and gluing graphs. Each graph can be constructed by gluing and identifying trees on their boundary points, so that a structure will be defined on certain labelled trees:

Definition 3.7 A rooted tree *is a labelled, directed (from bottom to top) trivalent tree, which means that the three edges at a branchpoint are labelled by objects and the starting root is labelled by the unit object.*

These objects have their origin in mod p representations of quantum groups, which provide *semisimple tensor categories* $\mathcal{T}$. We use three properties, with an additional requirement concerning the first property:

1. $\mathcal{T}$ consists of a finite set of simple objects; and the tensor product of two simple objects is a finite sum of simple objects. In addition, for each triple a_i, a_j, a_k of simple objects we demand: dim $(hom(a_i, a_j \otimes a_k)) = 1$ or $= 0$.
2. $\hom(a_i, a_j) = \delta_{ij} R$ for simple objects a_i, a_j and $R = \hom(1, 1)$, where 1 is the unit object in the tensor product.
3. For each simple object a_i there exists a unique *dual object* $\overline{a_i}$ determined by the unique appearance of the object 1 in the sum decomposition of $a_i \otimes \overline{a_i}$.

The example $\mathcal{T} = \mathfrak{sl}(2) \mod \mathbb{Z}_5$, $\mathcal{T}$ is generated by the simple objects $\{1, A\}$ with rules: $1 \otimes 1 = 1$, $1 \otimes A = A \otimes 1 = A$, $A \otimes A = 1 \oplus A \Rightarrow A = \bar{A}$, the unique dual object. Associativity and commutativity isomorphisms are defined by general identities (pentagon, triangle) in tensor categories, which describe their isomorphisms through their sum decompositions. The rooted tree defines a homomorphism of the unit object 1 into an ordered bracketing of tensor products of simple objects in the tensor category $\mathcal{T}$, depending on its labelling and geometric form. Reading from bottom to top, the homomorphisms at the *branch points* have to be composed. Let the edge under a branchpoint, the left departing edge and the right departing edge be labelled by x, y and z respectively. The corresponding homomorphism is defined by $\hom(x, y \otimes z) \neq 0$. Then, properties 1 and 2 imply that $y \otimes z$ contains at least a summand x. For the composition, each branch point has to be substituted by the tensor product $\otimes$.

Example 3.8 *(see Figure 3.5.c)): The left subtree defines the homomorphism:* $A \to A \otimes A \to A \otimes (A \otimes A)$, *for the middle subtree:* $A \to 1 \otimes A \to (A \otimes A) \otimes A$ *and for the right subtree:* $A \to A \otimes A \to (A \otimes A) \otimes A$. *Moreover Figure 3.5.a) and Figure 3.5.b) show further examples for different bracketings; the result is placed on the roots. With exception of the root, all edges are labelled by A. Note, that at the first branchpoint we have the condition* $\hom(1, A \otimes A) \neq 0$ *and for all others we have* $\hom(A, A \otimes A) \neq 0$. *Since* $A \otimes A = 1 \oplus A$, *both conditions are fulfilled by property 2.*

A rooted tree can be manipulated by sliding an arc over a branch point or by switching neighbouring departed edges on a common branch point. These

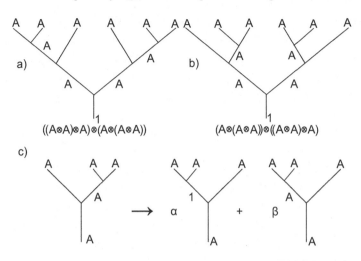

Figure 3.5 Different structures of rooted trees

elementary changes are related to structure maps in the tensor category $\mathcal{T}$, the associativity and commutativity isomorphisms. We illustrate an example in Figure 3.5.c), which shows the slide of an edge from right to left over a branch point. The associativity isomorphism is used to determine α and β. We refer to [Kad10] for this type of computations, see also [Tur94]. We present the gluing (cutting) of rooted trees. First we recall for a modular TQFT from Section 3.2, that identifying corner points in the boundary corresponds to identifying their associating ring structures on the state module of the boundary object. The boundary objects are the trees, the corner points are their boundary points. Note that the labelling is related to the edges. In Figure 3.6.a), we describe the gluing of two separated trees on their edges a, b: The rooted tree is manipulated by associativity and the applied non degenerate form λ; so for: $c \mapsto a \otimes b \stackrel{\lambda}{\mapsto} 1$ property 2 in $\mathcal{T}$ yields $c = 1$, hence $a = \bar{b}$. Figure 3.6.b) presents the special case of identifying two boundary points in a common rooted tree, say Y. The process is similar to the former one, first gluing intervals on the ends indicates the ring structure associated to the cornerpoints of the edges. In general left module structures (over the ring $Z(\overline{W} \times I)$) are preferred. We consider the left module structure of the edges x, y and convert it for one edge, say y, into a right module structure (over $Z(W \times I)$). The different ring structures of the corresponding corners are indicated by the labelling of the attached dotted intervals, see the second picture in Figure 3.6.b). We multiply both ring

Figure 3.6 gluing rooted trees

structures by gluing together these intervals in their version as rooted trees, see Figure 3.6.c). By the defining property 3 of the unique dual object, we obtain $\bar{y} = x$ or $\bar{x} = y$, namely $\bar{x} = y$ is the labelling in the third picture of Figure 3.6.b). The effect of the algebraic gluing of two corner points is fixed by labelling object and dual object as a pair. Thus we get a loop. This is expressed by the equivalence in Figure 3.6.b).

The 3-phases in the Quinn model - the algebraic context: For describing the start/end of the relator corresponding to the 2-cell, we need the expression for the bordism from $\emptyset$ to S^1 (and reverse), so that a bordism is a disc. The corresponding element is called the *trace unit* (see [Kad10]), its existence defines so called *special ambialgebras*. Finally, we arrange the main step of the *circulator* passing a corner, see Figure 3.7.a). First the generator defines a cut into a (rooted) tree, say Y with labelled edges (using the former result), say a and $\bar{a}$. The relator arc, labelled by r, where r is evaluated by the element of the trace unit, is not permitted to pass the endpoints. The idea is to simulate the passing. We introduce a second corner x, see the third picture in Figure 3.7.a). Suppose the relator arc r has passed x; then it has also passed the pair x, $\bar{x}$, obtained by cutting on x. Thus by gluing a, $\bar{a}$ together, the relator arc slides over the remaining part of the cutting generator (see the last picture in Figure 3.7.a)).

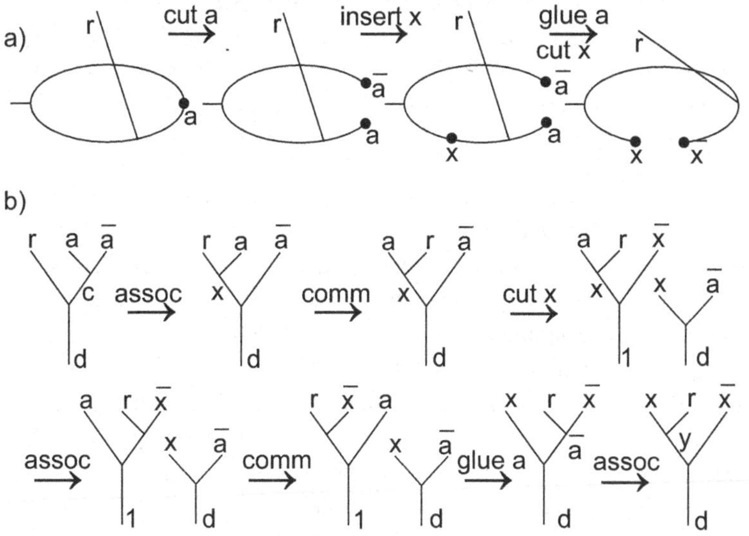

Aspects of TQFT and Computational Algebra 47

Figure 3.7 The circulator

This central trick is included in Figure 3.7.b). Note, that we have to consider all simple objects x compatible with the condition for branch points.

Definition 3.9 *The* circulator *in the Quinn model is defined by the sequence of rooted trees presented in Figure 3.7.b).*

Remark 3.10 *Each elementary transition between two neighbouring rooted trees can be described by the effect on the structure maps associativity, commutativity or gluing/cutting. Note that this is only a subtree, which has to be completed according to further generators and by the splitting counterpart $\bar{r}$ of the relator arc r, which slides only over the connecting rectangles between the generator cylinders (see the Quinn model, in Section 3.3.1).*

We will not define the state module $Z(Y)$ of a rooted tree, which uses the sum of all simple objects in $\mathcal{T}$. For our purpose to compute the invariant it is quite appropriate to give the definition of its base:

Definition 3.11 *Let Y_i be a rooted tree with the same geometric form of Y, labelled with simple objects which satisfy the condition on hom for all branch-points, then its defining homomorphism is a base element of $Z(Y)$.*

Remark 3.12 *Consider a subtree in a rooted tree Y. The effect of sliding or switching arcs yields a linear combination of the base, which corresponds to* subtrees. The coefficients are determined by the associated structure maps according to the change of the arcs, see example 3.8, Figure 3.5.c). By attaching the unchanged parts of Y to this base of subtrees, we obtain the base extension for rooted trees of the linear combination mentioned above. Applying the hom function this is also valid for their associated state modules.

So the circulator can be described as a matrix between bases, composed by the effect of its intermediate steps in the sequence. The circulator for an arbitrary generator can be obtained by switching the generators, application of the circulator and switching the generators back. Switching is supported by the structure map for commutativity. Supposing we have evaluated the matrices for start/end of relation /presentation (which requires further structure maps of the ambialgebra), we compose the matrices according to the slicing in the Quinn model (by the second TQFT axiom, see Section 3.2), which yields the Andrews-Curtis invariant on sliced 2-complexes due to Frank Quinn (in short *Quinn invariant*), in our example valuated in $\mathbb{Z}_5$. Note, that we have used gluing (cutting) of corner points on the boundary objects, the rooted trees, so the Quinn invariant is described by a modular TQFT.

3.3.3 Results on the Quinn invariant

I. Bobtcheva [Bob00] proves that the Quinn invariant defines an Andrews-Curtis invariant. She creates an invariant on the presentation of a 2-complex and shows it stays unchanged under Q^{**}-*transformations*. We list the topological basics of its construction (see [Man80]) by considering the 2-dimensional thickening of a 2-complex; a basepoint thickens to B^4, the generators to 1-handles $D^1 \times B^3$ with attaching maps $S^0 \times B^3$ into S^3; therefore two disjoint balls in S^3 have to be identified. Each relator thickens to a 2-handle $D^2 \times D^2$ with attaching map $S^1 \times D^2 \to S^3 \cup$ 1-handles, which simplifies to an attaching map $S^1 \times D^2 \to S^3$; this can be done by surgery on S^3; remove $S^1 \times D^2$ and attach a twisted $S^1 \times D^2$. Let $x \in D^2 \setminus \{0\}$; then $f(S^1 \times \{x\})$ wraps around $f(S^1 \times \{0\})$ n-times; the disc twists n-times around $f(S^1 \times \{0\})$; thus n is the linking number and the attaching leads to a link diagram, see for example the relation x^3y^2, $n = 9$, in Figure 3.8.a). The definition of the invariant includes a composition of these $x_i^{l_i}$, $l_i \in \mathbb{Z}$. For $l_i > 0$ it is close to the linking diagram of the relation, restricted to generators also having positive exponents. The x_k are presented as pairs of two boxes k, k^* (base and dual base due to a simple object k in a semisimple tensor category), which corresponds to cutting the 1-handle into two disjoint balls. If $x_k^{l_k}$, $l_k < 0$, then use the duality construction $\widehat{f}$ for a map f (use for f the case $l_k > 0$) for defining $x_k^{l_k}$ as indicated in Figure 3.8.b). *Bobtcheva's invariant* gets its data from

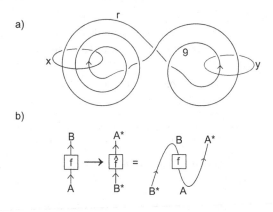

Figure 3.8 Link diagram of a relation and the duality morphism

a *coupon category*, factorized by substituting coupons with the identity on the strands. The coupon category is associated to a *stable, autonomous, tortile, semisimple tensor category*. As indicated above, the Bobtcheva invariant is defined on a 2-complex presentation, expressed in labelled link diagrams (with coupons). The category provides identities, which can be also described in labelled link diagrams with coupons, (see [FrYe89]). To verify the Andrews-Curtis-invariance of Bobtcheva's invariant, use these identities for manipulating the link diagrams into equivalent ones. The identities correspond to the Reidemeister-moves, to the identities for dual bases and to identities for the rank of an object and morphism. Here the property 2) of $\mathcal{T}$ will be used. By translating the circulator also into such a link diagram, it can be shown that the homomorphism reading a generator in Bobtcheva's invariant equals the circulator; so Bobtcheva's invariant equals the Quinn invariant. For the next results, see also Chapter 2, Section 2: K. Müller considers the TQFT of a presentation of the 2-complex as a TQFT group presentation of the 2-complex with generators; the circulators and relations are also expressed in terms of circulators due to the appearance of the generators in the relation. By Quinn's construction mentioned above all circulators have the same order. Therefore together with classical results many classes of groups (solvable, perfect, simple) for detecting Andrews-Curtis counterexamples are excluded. He also compares the TQFT group representation on two generators with the torus knot group and works out a further exclusion criterion for the abelianized TQFT group by using a result of Wes Browning about abelianized groups. Close to the one of Browning, in [BoLuMy05] there is presented a result due to the Andrews-Curtis graph of the finite test group and its abelianized one. It follows, that for the Andrews-Curtis conjecture (AC) there exists no finite test group and therefore no TQFT group with $Z(\emptyset) = \mathbb{Z}_p$, p prime, detecting counterexamples. Note, by the generalization of the result in [BoLuMy05] this is also valid for (AC'). In [BoQu05] the authors consider quantum invariants for 2-complexes which are *reductions of modular invariants* of their 4-dimensional thickenings. The main result is:

Theorem 3.13 *For $\chi(K^2) \geq 1$ the reductions of these invariants depend only on homology. Furthermore, for $\chi(K^2) < 1$ the reduction of the $SO(3)$ quantum invariant at the 5-th root of unity is able to distinguish 2-complexes with the same homology groups.*

Remark 3.14 *For commutative rings R, K: Z_K^4 is multiplicative with respect to (boundary) connected sum, that is $Z_K^4(M_1 \#_\partial M_2) = Z_K^4(M_1) Z_K^4(M_2)$. Z_R^2 is called the R-reduction of Z_K^4. It is multiplicative with respect to one-point union of 2-complexes.*

Note, that there is no result for the Quinn invariant to be useless in the case of the relative Andrews-Curtis conjecture (rel AC').

3.3.4 The Multiplicative Property of Modular TQFT

The results in [BoQu05] imply, that modular invariants can not detect (AC) counterexamples. Hence for (AC) only the construction of non modular TQFT invariants seems promising. The next lemma facilitates a test for non modular TQFT invariants:

Lemma 3.15 *For 2-complexes K^2, L^2 and a modular TQFT Z, $Z(K^2 \vee L^2) = Z(K^2)Z(L^2)$ holds. Furthermore [Hu01] cites (without a reference) a multiplicative property on the connected sum of closed 3-manifolds M_1^3, M_2^3: $Z(M_1^3 \# M_2^3)Z(S^3) = Z(M_1^3)Z(M_2^3)$ for an arbitrary TQFT Z.*

Proof: We outline a proof of these statements, illustrated in Figure 3.9. For the statement on 2-complexes see Figure 3.9.a): We regard the 2-complexes K^2 and L^2 in the Quinn model, attached into different generator cylinders with a common point. This common point can be substituted (by 3-deformations and further 2-deformations) by connecting flanges. By definition a modular TQFT Z provides a unit $1 \in Z(W \times I)$, $Z(\partial Y \times I)$ (see Section 3.2), which acts by the identity on $Z(Y)$. Realize the connecting flanges, this forces us to use that unit;

Aspects of TQFT and Computational Algebra

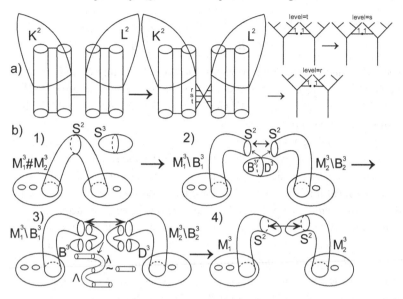

Figure 3.9 Multiplicative property of TQFT

choose arbitrary slices (in form of rooted trees) with, say, level $t < s < r$, and denote their state modules by $Z(Y_t)$, $Z(Y_s)$, $Z(Y_r)$. Add (on arbitrary places) arcs labelled with the unit 1 as indicated in Figure 3.9.a). For level t, r the arcs are disjoint, at level s they meet in a common corner. The added arcs with label 1 induce an action by the identity on $Z(Y_t)$, $Z(Y_s)$, $Z(Y_r)$; the homomorphisms between these state modules also stay unchanged. So we are exactly in the situation, where the 2-complexes K^2 and L^2 are disjoint, which implies the multiplicative property by the first TQFT axiom for disjoint bordisms, using $Z(K^2)$, $Z(L^2) \in R$ (see Section 3.2). For the statement on 3-manifolds we will use only TQFT axioms and the fact, that for a closed 3-manifold M^3, $Z(M^3) \in R$. We start with $Z(M_1^3 \# M_2^3) Z(S^3)$; this corresponds by the first TQFT axiom combined with the fact, that $M_1^3 \# M_2^3$, S^3 are closed manifolds, to the disjoint union of $M_1^3 \# M_2^3$ and S^3, depicted in Figure 3.9.b)1). The connected sum of $M_1^3 \# M_2^3$ is presented with a tube $S^2 \times I$ connecting the S^2 boundaries of the 3-balls B_1^3 in M_1^3 and B_2^3 in M_2^3 being removed. For the transition to Figure 3.9.b)2) we apply the second axiom and interpret the gluing on S^2 as a composition of $Z(M_2^3 \setminus B_2^3) \circ Z(M_1^3 \setminus B_1^3)$, indicating the identification on the S^2 boundaries by the left/right arrow. Since $Z(S^3) \in R$, the S^3 can be placed in the middle and we get:

$$Z(M_2^3 \backslash B_2^3) \circ Z(S^3) \circ Z(M_1^3 \backslash B_1^3)$$

The transition to Figure 3.9.b)3) yields, that we split S^3 into B^3 and D^3 and use the composition property in the second axiom; so we have to glue $Z(M_1^3 \backslash B_1^3)$ with B^3 and $Z(M_2^3 \backslash B_2^3)$ with D^3. However gluing outgoing boundaries is not permitted. For each tube we convert its outgoing S^2 boundary into an incoming boundary, using a further tube with turn to B^3 respectively D^3. The attached tubes to B^3, respectively D^3, define two 3-balls. By the former identification of the S^2 boundaries we also get a 3-sphere S^3, and not two separated 3-balls!! Note, the resulting bordisms M_1^3 and M_2^3 have additional entries coming from B^3, D^3. That requires a little modification: Applying again a further tube with turn in opposite direction provides bordisms equivalent to the cylinder by the collar axiom, see the added little picture in Figure 3.9.b)3; hence the associated homomorphisms, called *dual pairing* by (λ, Λ) are the identity. Again the collar axiom provides bordisms (the pair of tubes is substituted by a cylinder), equivalent to M_1^3, M_2^3, as depicted in Figure 3.9.b)4. Since M_1^3 and M_2^3 are closed, the composition $Z(M_2^3) \circ Z(M_1^3)$ equals the multiplication $Z(M_1^3) Z(M_2^3)$ in R. The union of the resulting bordisms is $M_1^3 \cup_{S^2} M_2^3$, illustrated by the left/right arrow in Figure 3.9.b)4). We use only the TQFT axioms, so the statement holds also for $n \geq 3$. □

3.4 TQFT on s-move 3-cells

In [I], [Qu85] Quinn presents the topological exploration, that two simple homotopy-equivalent (in short: sh-equivalent) 2-complexes are related by an s-move. This can be combined with the corresponding commutator criterion for two sh-equivalent 2-complexes given in [I] or Chapter 2, Section 2, for potential constructions of Andrews-Curtis invariants on sliced s-move 3-cells. We provide the underlying facts and refer to [Kad17] for details.

3.4.1 Basics about the s-move

Data of the s-move: Let K^2 and L^2 are two 2-complexes with a common 1-skeleton. Let $P(K^2) = \langle a_i \mid R_* \rangle$ and $P(L^2) = \langle a_i \mid S_* \rangle$. The data for the s-move between K^2 and L^2 are oriented surfaces with attached annuli (indicated by directional arcs), depicted in Figure 3.10.a):

Aspects of TQFT and Computational Algebra 53

- Each meridian curve is connected via an annulus to the boundary of a disc, which will be mapped to an S_* relator, and each longitudinal curve is connected to the boundary of a disc, which will be mapped to an R_* relator.
- Fold collars of the generator curves of the surfaces to create the annuli and attach these to the boundaries of the labelled discs.

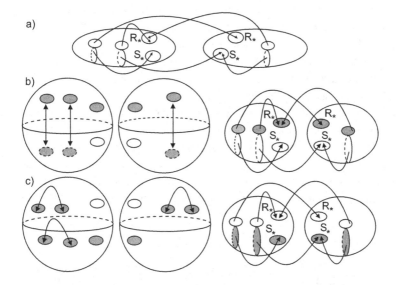

Figure 3.10 s-move data and the elementary 3-expansion to s-move 3-cells

Definition 3.16 *In the situation above, we say that K^2 and L^2 are related by an s-move.*

Remark 3.17 *Data of the s-move are the perforated surfaces with holes R_*, S_* together with the attached annuli which are mapped into the (common) 1-skeleton.*

The s-move 3-cells as an elementary 3-expansion: K^2 and L^2 can be extended (one for each pair R_*, S_* on the S^2) by an elementary 3-expansion of a 3-cell $\subset K^3$, $\subset L^3$ with free faces S_*, R_*, see Figure 3.10.b) for K^2 and Figure 3.10.c) for L^2: Identify subdiscs on the 2-spheres with holes; this gives oriented surfaces with holes and attached discs (the identified subdiscs) for the half set of generators (longitudinals for K^2, meridians for L^2). Pick up smaller discs and attach these to discs labelled R_*, S_* for K^2, L^2. Filling 3-balls into each 2-sphere fills the holes S_* in K^2 and R_* in L^2 with discs. The filling discs are the free cells; the filling discs on the end of the annuli are not attached. Following the remark, we have to perform this construction in the image.

Theorem 3.18 *The s-move 3-cells K^3 and L^3 are sh-equivalent*

Proof: By Lemma 2.1 of [I] it is sufficient, that the attaching maps of related 3-cells (via the pair R_*, S_* for a fixed index $*$) are homotopic. For the proof we follow Quinn, see [I], [Qu85]: The idea is to simplify the homotopy of the attaching maps by a pullback of its image. We consider the homotopy for the spheres themselves in a simplified situation: Figure 3.11.a) shows the sequence; the start sphere in the first picture, with identified subdiscs attached to the meridian. Take a neighbourhood of the (whole) longitude and push it across the longitudinal disc and the meridian disc of a fixed torus, with attached discs for both generators. In the second picture, the 2-material slides between the identified subdiscs (denote by S_α) and therefore separates these. In the last picture we get a 2-sphere (having embedded separated discs S_α) with identified subdiscs attached to the longitude, hence a torus with an attached disc on the longitude, as required. That disc is generated by crossing the longitudinal disc. □

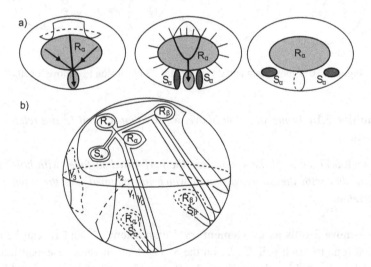

Figure 3.11 s-move 3-cells: homotopy of attaching map and association to the commutator criterion

3.4.2 The decomposition of the s-move 3-cell into 2-cells

The equivalence of the s-move construction with the commutator criteria for sh-equivalent 2-complexes: We use the commutator criterion of Chapter 2, Section 2 of this book: If K^2 and L^2 are sh-equivalent, then we consider the simplified case

*) $\quad R_* S_*^{-1} = [R_\beta, S_\beta][R_\alpha, S_\alpha]$ respectively

**) $\quad R_* S_*^{-1}[S_\alpha, R_\alpha][S_\beta, R_\beta] = 1$

for a relator pair R_*, S_*. The latter is more suitable in our situation; it is an equation of the trivial word in the free group $F(a_i)$. Elementary cancellations applied to the left side of this equation describe a homotopy for a map into the common 1-skeleton of $K^2 \cup L^2$ to the constant map into the base point P. Figure 3.11.b) shows the pullback of this homotopy to a 1–parameter family of graphs γ_t with $t \geq 0$ on the 2-sphere, indicated by a sequence of graphs $\gamma_0, \gamma_1, \gamma_2$ and γ_3 which shrinks to a point. The start graph γ_0 is mapped to the word corresponding to **). The holes will be filled by discs, mapped to the relators according to their label. Identifying the subdiscs in the 2-sphere according to their label results in an orientable surface with discs attached to the half set of generators. These are the longitudes R_α, R_β; the other generators S_α, S_β identify to meridian curves. The case for K^3 is depicted in Figure 3.11.b). This confirms the mapping of the s-move data into the 1-skeleton. Note, we can assume from the commutator criterion the relators R_α, R_β, S_α, S_β in a conjugation form which indicates the annuli.

The significant neighbouring slices of the s-move 3-cell: The 1–parameter family of graphs γ_t on the 2-sphere, depicted in Figure 3.11.b), can be extended to a characteristic map on the 3-ball, see the left pictures in Figure 3.12). We attach a disc to a circle, the image in $K^2 \cup L^2$ is the 2-cell according to the label of the circle. We attach a bag for an edge with label X, mapped to a 2-cell with relation XX^{-1}. Non labelled bags are attached to arcs, mapped to the base point. They serve only for a simplified drawing. Figure 3.12 describes the essential difference between two s-move 3-cells ; Figure 3.12.a) does the same for a *longitudinal identification* and Figure 3.12.b) for a meridian. Both types (longitudinal or meridian) have the same common (commutator-) 2-cells (see the right pictures in Figure 3.12) but different squeezed (commutator-) 2-cells (see the left pictures in Figure 3.12). This may help to distinguish them. It is the central motivation of these notes to include this in a potential Andrews-Curtis invariant on s-move 3-cells. Note, that by the identification of the subdiscs on S^2 the boundaries of the pairs R_α and R_β in the Figure 3.12.a) and S_α and S_β

in Figure 3.12.b) are also being identified; they are called *spherical elements* (together with the bags).

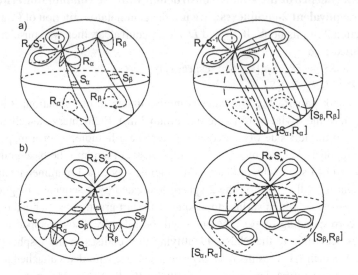

Figure 3.12 distinguish s-move 3-cells by squeezed commutator 2-cells

3.4.3 Considerations for constructing an Andrews-Curtis invariant

Independence of the type of identification and invariance inside an identification type: First we present the essential steps in the slicing of an s-move 3-cell $\subset K^3$ in Figure 3.13; Figure 3.13.1) changes into Figure 3.13.2) by joining R_* and S_* to $R_*S_*^{-1}$. Figure 3.13.2) becomes Figure 3.13.3) by changing the squeezed commutator 2-cells to the ordinary commutator 2-cells. Figure 3.13.3) gets to Figure 3.13.4) by joining the separated 2-cell pieces to a single 2-cell with trivial boundary, since it presents the graph of the commutator criterion. Similarly we can perform the slicing of an s-move 3-cell $\subset K^3$. For the invariance of the identification type we can use $K^2 = L^2$, hence for a pair R_*, S_* in our simplified commutator criterion $*)R_*S_*^{-1} = [R_\beta, S_\beta][R_\alpha, S_\alpha]$ we can take: $R_* = S_*, R_\alpha = S_\alpha$ and $R_\beta = S_\beta$. This induces the same labelled 2-cells and hence the same slice. Apply this to the construction of the potential invariant. For the Andrews-Curtis invariance inside an identification type, we choose the longitudinal identification for K^3 and consider the *Q-transformations*, see [I]. The sequence of Q-transformations $(K^2, L^2) \to (K'^2, L^2) \to (K'^2, L'^2)$

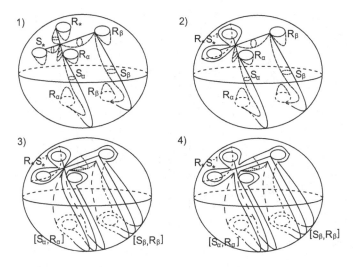

Figure 3.13 the slicing of the s-move 3-cell for longitudinal identification

shows, that we can study all of them in (K^2, L^2) by restrictions to one set of relators, say $\{R_*\}$ in K^2, keeping fixed the $\{S_*\}$ and vice versa. Let us assume the simplified commutator criterion *). For the application of a Q-transformation $R_* \to R'_*$, we can arrange, that $R'_* S_*^{-1} = \tilde{R}_*^{-1}[R_\beta, S_\beta][R_\alpha, S_\alpha]$; similarly we get for a Q-transformation $S_* \to S'_*$, that $R_* S'^{-1}_* = [R_\beta, S_\beta][R_\alpha, S_\alpha]\tilde{S}_*^{-1}$. Here is an easy example: $R'_* = R_*^{-1} \Rightarrow R'_* S_*^{-1} = R_*^{-1} R_*^{-1} R_* S_*^{-1} \Rightarrow \tilde{R}_*^{-1} = R_*^{-1} R_*^{-1}$. For the Q-transformation $R'_* = R_* R_k$ use $[a, b]ba = ab$. We interpret $\tilde{R}_*$ as a further hole in the s-move data due to the unchanged orientable surface (commutator product), being capped by 2-cells in K^2 (K'^2) with $\tilde{R}_* R'_* = R_*$; treat for $\tilde{S}_*$ analogously.

The abstracted model of a sliced s-move 3-cell: We simplify the drawing by an abstraction of the sliced s-move 3-cell, see Figure 3.14.a) without Q-transformations and Figure 3.14.b), when a Q-transformation was performed. Note, that we have omitted the first and last slice; the empty set which is related to the 2-cells; thus the slice is the wedge of generator cylinders (see the Quinn model, in Section 3.3.1). We denote the spherical elements by SP-EL. Using the condition $\tilde{R}_* R'_* = R_*$, we compare both slicings. We almost get the same slices (because of equal boundary words) with one exception, namely the second slice with the separated 2-cells $\tilde{R}_*$, R'_* in Figure 3.14.b). This forces us to find a way defining the invariant insensible to the intermediate slice. The situation at least indicates a composition property for the case in question. These

Figure 3.14 the abstracted model of a sliced s-move 3-cell

considerations complete our survey.

3.5 A general construction of state sum invariants

Many well known TQFT-invariants, such as Turaev–Viro invariants of 3-manifolds [Tur94, Chapter VII], or the Jones–Kauffman polynomial of links [Kau87, Chapter IV], are associated with a "finite calculus": Topological data are presented by means of combinatorial data and combinatorial data presenting essentially the same topological data can be related by finitely many types of local transformations. The aim of this section is to provide a general framework for the construction of "state sum invariants" taking values in commutative rings, for any given finite calculus.

3.5.1 What do we mean by a *finite calculus*?

In a finite calculus, the objects of interest (say, knots, 3–manifolds, 4–thickenings) are encoded by topological spaces that are built upon a finite number of blocks, and the number of available block types is finite. This can be made precise as follows. We do apologize for introducing a rather heavy notation for an idea that is intuitively clear, for the sake of generality.

Aspects of TQFT and Computational Algebra 59

Definition 3.19 *Let $U_1, \ldots, U_k$ be a finite list of topological spaces. Let X be a topological space.*

1 If $x \in X$ has a neighbourhood $U_x \subset X$ homeomorphic to U_i (for some $i = 1, \ldots, k$) such that U_x does not contain a neighbourhood of x homeomorphic to U_j with $j = 1, \ldots, k$ and $j \neq i$, then we say x is of local type U_i, and U_x is a typical neighbourhood *of x.*

2 For $i = 1, \ldots, k$, a connected component of $\{x \in X \mid x \text{ is of local type } U_i\}$ is called a U_i–stratum of X. We denote the set of all strata (of either local type) of X by $\Theta(X)$.

3 We call $U_1, \ldots, U_k$ local types, if for all $i = 1, \ldots, k$, for each $x \in U_i$ there is some $j = 1, \ldots, k$ so that x is of local type U_j.

4 If U_i is, for some space Z, homeomorphic to $Z \times]-1, 1[$, and $\overline{U}_i := Z \times]-1, 0]$ is not homeomorphic to any of $U_1, \ldots, U_k$, then $\overline{U}_i$ is called the boundary type *associated with U_i. As above, we also define points of local type $\overline{U}_i$, and $\overline{U}_i$-strata of X, which we call* boundary strata. *The union of boundary strata of X is denoted by ∂X. Note that by definition no stratum in $\Theta(X)$ is a boundary stratum, but each boundary stratum is contained in the closure of some stratum in $\Theta(X)$.*

5 Let $U_1, \ldots, U_k$ be local types. We say that X is locally built *upon $U_1, \ldots, U_k$, if*

- *X is a finite union of U_i-strata (for $i = 1, \ldots k$) respectively $\overline{U}_i$-strata (for those $i = 1, \ldots, k$ for which the boundary stratum associated with U_i is defined), and*

- *if x, y belong to the same stratum $s \in \Theta(X)$ and U_x, U_y are typical neighbourhoods of x, y, then there is a homeomorphism $\phi: U_x \to U_y$ such that for all strata $t \in \Theta(X)$ holds $\phi(t \cap U_x) = t \cap U_y$. If $t \cap U_x \neq \emptyset$, we say that t is* adjacent *to s (which includes the case $t = s$).*

6 Let $\mathcal{G} = (G_1, \ldots, G_k)$ be commutative groups, and X a space that is locally built upon $U_1, \ldots, U_k$. A $\mathcal{G}$-decoration of a U_i–stratum $s \in \Theta(X)$ is an element of G_i. A $\mathcal{G}$-decoration δ of X assigns a $\mathcal{G}$-decoration $\delta(s)$ to each $s \in \Theta(X)$. We denote the group operations by $\cdot$.

Note that we do not necessarily assume that the strata of X are cells, although this is the case for many applications. Also note that any $\overline{U}_i$-stratum is contained in the closure of some U_i-stratum.

Example 3.20

1 A 3-valent graph is locally built upon $U_1 =]-1, 1[$ (open interval) and the open core of the cone over three points.

2. Let U_1 be an open disc, $U_2 = $ [figure] and $U_3 = $ [figure]. Here, both U_2 and U_3 are open, i.e., the boundary seen in the pictures does not belong to the neighbourhood. A simple 2-polyhedron is built upon U_1, U_2, U_3. Note that U_2 is of the form $Z\times]0, 1[$, where Z has the shape of the letter Y (three half-open edges joined in one point). So, the associated boundary type $\overline{U}_2$ looks like the picture for U_2 except that one Y-shaped graph in the visible boundary actually is contained in $\overline{U}_2$ (but not in U_2).

3. In some cases, we focus on special polyhedra, which are simple 2-polyhedra without boundary for which the U_1-strata (i.e., strata that are surfaces) are discs, and the union of U_2- and U_3-strata is a connected graph with at least two vertices. For any closed 3-dimensional manifold M there is an embedded special polyhedron $P \subset M$ such that $M \setminus P \approx B^3$. In that case, P is called a [1] special spine of M. The homeomorphism type of M is uniquely determined by the homeomorphism type of [1], [Ca65].

4. 4-dimensional thickenings (see Section 3.3.3 or [Tur94]) of special 2-polyhedra are determined by the assignment of an integer or half-integer to each U_1-stratum. Hence, it is determined by a $(\frac{1}{2}\mathbb{Z}, 1, 1)$-decoration. We give more details in Section 3.5.2 below.

Definition 3.21 Let $U_1, \ldots, U_k$ be local types and $\mathcal{G} = (G_1, \ldots, G_k)$ be commutative groups.

1. A local move type is a 5-tupel $T = (A_1, A_2, \Gamma_1, \Gamma_2, \phi)$, such that for $i = 1, 2$ we have A_i, Γ_i compact and connected, $\Gamma_i \subset A_i$, Γ_i is contained in the closure of $A_i \setminus \Gamma_i$, $A_i \setminus \Gamma_i$ is locally built upon $U_1, \ldots, U_k$ and $\phi \colon \Gamma_1 \to \Gamma_2$ is a homeomorphism.

   If the closure $\bar{s} \subset A_i$ of a stratum s of $A_i \setminus \Gamma_i$ is disjoint from Γ_i ($i = 1, 2$), then we call s an inner stratum of A_i. Otherwise, we call it an outer stratum of A_i. If s_1, s_2 are outer strata of A_1, A_2 of the same type, and $\phi(\overline{s_1} \cap \Gamma_1) = \overline{s_2} \cap \Gamma_2$, then s_1, s_2 are corresponding strata of T.

   T is $\mathcal{G}$-decorated if a $\mathcal{G}$-decoration for the inner strata of A_1 and A_2 and for the outer strata of A_2 are chosen.

2. Let X_1, X_2 be locally built upon $U_1, \ldots, U_k$ and let $T = (A_1, A_2, \Gamma_1, \Gamma_2, \phi)$ be a local move type. We say that X_2 is obtained by applying a move of type T to X_1, if

   - for $i = 1, 2$ there are embeddings $\iota_i \colon A_i \hookrightarrow X_i$ and a homeomorphism $\psi \colon X_1 \setminus \iota_1(A_1 \setminus \Gamma_1) \to X_2 \setminus \iota_2(A_2 \setminus \Gamma_2)$, such that $\psi \circ \iota_1|_{\Gamma_1} = \iota_2 \circ \phi$,
   - for $i = 1, 2$ the local type of any $x \in A_i \setminus \Gamma_i$ coincides with the local type of $\iota_i(x) \in X_i$,

Aspects of TQFT and Computational Algebra   61

- *for any $x \in \Gamma_1$, the local type of $\iota_1(x) \in X_1$ coincides with the local type of $\iota_2(\phi(x)) \in X_2$, and*
- *if ι_1 maps an outer stratum $\tilde{s}_1$ of A_1 into a stratum s_1 of X_1, then there is a unique stratum s_2 of X_2 such that ι_2 maps a stratum of A_2 into s_2 that corresponds to $\tilde{s}_1$. In that case, s_1, s_2 are* corresponding strata *of the move.*

3 *If both X_1, X_2 and T are $\mathcal{G}$-decorated, then it is additionally assumed that*
- *if s is a stratum of X_1 disjoint from $\iota_1(A_1)$ then s and $\psi(s)$ (as a stratum of X_2) have the same $\mathcal{G}$-decoration,*
- *if s is a stratum of A_1 disjoint from Γ_1 (resp. of A_2 disjoint from Γ_2), then s and the stratum $\iota_1(s)$ of X_1 (resp. $\iota_2(s)$ of X_2)) have the same $\mathcal{G}$-decoration, and*
- *if s_1, s_2 are corresponding strata of the move, $g_1, g_2 \in G_j$ (for some $j = 1, \ldots, k$) are the $\mathcal{G}$-decorations of s_1, s_2, and g is the $\mathcal{G}$-decoration of the U_j-stratum $\iota_2^{-1}(s_2)$ of $A_2 \setminus \Gamma_2$, then $g_2 = g_1 \cdot g$.*

Note that by our definition, local moves leave the boundary untouched.

Definition 3.22 *Let $U_1, \ldots, U_k$ be local types and $T_1, \ldots, T_l$ be local move types. A* finite calculus *is a (multi-)graph, whose vertices are given by homeomorphism types of compact spaces that are built upon $U_1, \ldots, U_k$, so that there is an edge joining a vertex X_1 with X_2 if and only if there is a local move of type T_j (for some $j = 1, \ldots, l$) relating X_1 and X_2 (in either direction).*

Of particular interest is to provide means to determine whether or not two vertices of a finite calculus belong to the same connected component.

3.5.2 Examples of local calculi

1 By [I], [Ma87$_1$] and [I], [Pi88], if P_1, P_2 are special spines of the same closed 3-manifold M, then P_1 and P_2 are related by finite sequences of the local move depicted in Figure 3.15; see also Figure 3.4 for a different depiction of the move. The move is called *T move* in [Mat03] and is also called Matveev–Piergallini move, see [I], [Pi88]. In the figure, A_1 is left and A_2 is right, and Γ_i is the visible boundary of A_i. To make the picture easier to understand, points of type U_3 are marked by a thick dot. Note that the triangle in the middle of A_2 is the boundary of a U_1-stratum.
2 We give a brief outline of results on ($\frac{1}{2}\mathbb{Z}, 1, 1$)-decorated simple polyhedra in [Tur94, Chapter VIII and IX]. We specialize here to *special* polyhedra, i.e., to simple polyhedra without boundary whose 2-strata are homeomorphic to discs, having at least two U_3-strata, and so that the union of U_2– and

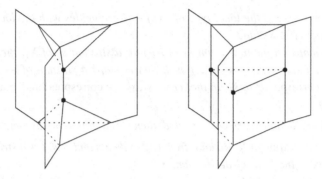

Figure 3.15 The T move

U_3-strata forms a connected graph. The boundary curve of a 2-stratum s of a special polyhedron X has a regular neighbourhood in X that may or may not contain a Möbius strip whose core is the boundary of s. If the regular neighbourhood does not contain such a Möbius strip, then the decoration of s is an integer. Otherwise, the decoration of s is a half-integer (that's to say, $\frac{n}{2}$ for an odd number $n \in \mathbb{Z}$). The decoration (which is called *gleam* in [Tur94]) defines a 4-dimensional thickening of X. Let $X^{(1)}$ denote the union of points of type U_2 and U_3 (i.e., the complement of the 2-strata of X). If X is embedded in two 4-dimensional manifolds W_1, W_2, then regular neighbourhoods of $X^{(1)}$ in W_1 and W_2 are homeomorphic. However, regular neighbourhoods of X in W_1 and W_2 need not to be homeomorphic, as the self-intersection numbers of any 2-stratum s of X may be different in W_1 and in W_2. Indeed, the regular neighbourhood of s in W_i ($i = 1, 2$) is homeomorphic to $D^2 \times D^2$, and is glued to the regular neighbourhood of $X^{(1)}$ along the solid torus $S^1 \times D^2$. The gluing map is determined up to isotopy by the linking number of the image of a longitude of the boundary torus and the image of the core of the solid torus, which corresponds to the integral part of the decoration.

Let X_1, X_2 two ($\frac{1}{2}\mathbb{Z}, 1, 1$)-decorated special polyhedra, whose decorations determine 4-dimensional thickenings. It follows from [Tur94, Chapter IX, Theorem 1.7] that on obtains from the two thickenings the same compact oriented 4-manifold by attaching 3- and 4-handles, if and only if X_1 and X_2 are related by the so-called *basic shadow moves* depicted in Figure 3.16 together with the bubble move depicted in Section 3.6.2. The numbers 0 and $\frac{1}{2}$ appearing in Figure 3.16 indicate the decoration of inner strata respectively the change of decoration of outer strata. Note that a third move appears

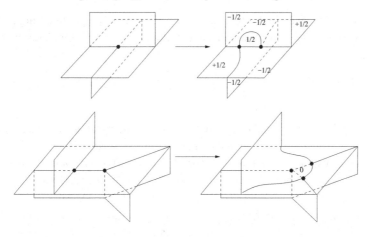

Figure 3.16 Basic shadow moves

in [Tur94], which in the case of *special* (as opposed to *simple*) polyhedra factors through the other shadow moves.

3 We remark that the Reidemeister calculus of link diagrams can be formulated in our setting as well, but we won't go into details.

3.5.3 State sums

Let $U_1, \ldots, U_k$ be local types upon which a space X is built, let $C_1, \ldots, C_k$ be finite sets, let $\mathcal{G} = (G_1, \ldots, G_k)$ be abelian groups and let R be a commutative ring.

Our aim is to describe a procedure to associate an R-module $\mathcal{Z}(C)$ to each connected component C of ∂X, and a so-called state sum to X, which is an element $\mathcal{Z}(X)$ of $\bigotimes_{C \text{ component of } \partial X} \mathcal{Z}(C)$.

Here is the outline. The basic data of a state sum construction consists of a choice of a finite set C_i of *colours* for each local type U_i and a choice of an element of R for each (ordered) tuple of colours. A *state* of X (or ∂X) consists of a choice of a colour for each stratum of X (or ∂X). Now, a basis of $\mathcal{Z}(\partial X)$ is given by the states of ∂X, and the construction of $\mathcal{Z}(X)$ relies on the states of X that restrict to a given state of ∂X. Each stratum s of X is adjacent to finitely many strata of X, whose colours form a tuple associated with s, which in turn determine an element of R. We multiply these elements, for all strata of X, and obtain an element of R. In the end, we form the sum of these elements over all states of X restricting to a given state of ∂X, which is the *state sum* of X.

It should be noted that the state sum will generally change when applying a local move; however, in the next section, we construct a polynomial ring P and an ideal $I_T \subset P$ such that the state sum modulo I_T is invariant under moves of type T.

Definition 3.23

1. *A colour of a U_i–stratum of X is an element of C_i.*
2. *A state Φ of X is a choice of a colour $\Phi(s)$ for each stratum $s \in \Theta(X)$. Similarly, a state of ∂X is a choice of a colour in C_i for each $\overline{U}_i$-type boundary stratum. A state Φ of X induces a state $\Phi|_{\partial X}$ of ∂X, as each boundary stratum of type $\overline{U}_i$ is contained in the closure of a unique U_i-stratum of X.*
3. *The set of the states of X (or ∂X) is denoted by $\Sigma(X)$ (or $\Sigma(\partial X)$).*
4. *Two states $\Phi_1, \Phi_2 \in \Sigma(X)$ are equivalent, if there is a homeomorphism $\phi \colon X \to X$ such that for all $s \in \Theta(X)$, $\Phi_1(s) = \Phi_2(\phi(s))$ (note that since ϕ is a homeomorphism, $\phi(s)$ is a stratum of X if s is a stratum of X).*

Remark 3.24 *One can modify the notion of equivalence of states by taking into account additional data. In [Kin07a], King studies several constructions of state sum invariants for special polyhedra, and in some of them, orientations of U_1-strata are taken into account in the local moves. Here, C_1 is endowed with an involution, and changing the orientation of a coloured U_1-stratum corresponds to applying the involution to its colour. The equivalence of states then has to comply with orientations as well.*

Definition 3.25 *Let U_i be a local type. A U_i–symbol is an equivalence class of states of U_i.*

Definition 3.26 *Let $\Phi \in \Sigma(X)$, let s be a U_i–stratum of X, and $x \in s$ with a typical neighbourhood $U_x \in X$. The colouring Φ of X restricts to a state of U_x, which gives rise to a U_i–symbol denoted by s_Φ.*

Note that s_Φ is well-defined, *i.e.*, independent of the choice of $x \in s$, since the typical neighbourhoods of any two elements of s are related by homeomorphisms that are compatible with the neighbourhood's intersection with the strata of X.

For each $i = 1, \ldots, k$, fix a choice of an element $|\sigma| \in R$, for each U_i–symbol σ, and for each colour $c \in C_i$ a homomorphism of multiplicative monoids $\epsilon_c \colon G_i \to R$ (for simplicity, we assume that the $C_1, \ldots, C_k$ are pairwise disjoint, so that we do not need to include i in the notation of ϵ_c).

Definition 3.27 *Fix a $\mathcal{G}$-decoration δ on X.*

1. *If Φ is a state of X, then let $|X|_\Phi := \prod_{s \in \Theta(X)} \left(|s_\Phi| \cdot \epsilon_{\Phi(s)}(\delta(s)) \right)$.*

2 If a state $\phi \in \Sigma(\partial X)$ is fixed, then the state sum of X with respect to ϕ is $\|X\|_\phi := \sum_{\Phi \in \Sigma(X), \phi = \Phi|_{\partial X}} |X|_\Phi$.

Example 3.28 *We recall here the symmetries of the basic data for the construction of the Tureaev–Viro invariants from [TuVi92]. We consider simple polyhedra, and choose $C_2 = C_3 = \{0\}$. Hence, the symbol associated with a U_1-stratum is determined by its colour $c \in C_1$; we denote it by $|c|$. The symbol associated with a U_2-stratum is determined by the set $\{c_1, c_2, c_3\}$ of colours of the three adjacent U_1-strata; we call it 3j-symbol and denote it by $|c_1\ c_2\ c_3|$. Note that $|c_1\ c_2\ c_3| = |c_2\ c_1\ c_3| = |c_2\ c_3\ c_1|$. The symbol associated with a U_3-stratum is determined by the colours of the six adjacent U_1-strata, as shown here*

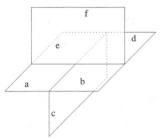

which is also called the 6j-symbol $\left| \begin{smallmatrix} a & b & c \\ d & e & f \end{smallmatrix} \right|$. By the tetrahedral symmetry of U_3, we have $\left| \begin{smallmatrix} a & b & c \\ d & e & f \end{smallmatrix} \right| = \left| \begin{smallmatrix} e & d & c \\ b & a & f \end{smallmatrix} \right| = \left| \begin{smallmatrix} c & e & d \\ f & b & a \end{smallmatrix} \right|$.

Choosing $C_1 = \{1, 2\}$, one obtains eleven 6j-symbols:

$$\left|\begin{smallmatrix}1&1&1\\1&1&1\\2&2&1\\2&2&1\end{smallmatrix}\right|, \left|\begin{smallmatrix}2&1&1\\1&1&1\\2&2&2\\1&1&1\end{smallmatrix}\right|, \left|\begin{smallmatrix}2&1&1\\2&1&1\\2&2&2\\2&1&1\end{smallmatrix}\right|, \left|\begin{smallmatrix}2&2&1\\1&1&1\\2&2&2\\2&2&1\end{smallmatrix}\right|, \left|\begin{smallmatrix}2&2&1\\1&1&2\\2&2&2\\2&2&2\end{smallmatrix}\right|, \left|\begin{smallmatrix}2&2&1\\2&1&1\\\end{smallmatrix}\right|.$$

3.5.4 Reading off the invariance conditions of a finite calculus

In this section, we consider a finite calculus given by local types $U_1, \ldots, U_k$ and local move types $T_1, \ldots, T_l$, we fix abelian groups $\mathcal{G} = (G_1, \ldots, G_k)$ and pairwise disjoint finite sets $C_1, \ldots, C_k$ of colours. Additionally, we fix a commutative ring R_0 and, for each $c \in \bigcup_{i=1}^{k} C_i$, a multiplicative homomorphism $\epsilon_c: G_i \to R_0$, where i is uniquely determined by c so that $c \in C_i$. We also fix a state $\phi \in \Sigma(\partial X)$.

It is obvious from the definition that the state sum of a $\mathcal{G}$-decorated topological space X built upon $U_1, \ldots, U_k$ will generally change when applying one of the moves $T_1, \ldots, T_l$. However, it is generally possible to choose a ring R containing R_0 as a sub-ring and choose $|\sigma| \in R$ for U_i-symbols ($i = 1, \ldots, k$), so that $\|X\|_\phi$ only depends on the component of the finite calculus which X

belongs to: Trivially, one can always take $R = R_0$ and send all symbols to zero. The aim of this section is to describe a general construction of less trivial state sum invariants.

Let P be the *multivariate polynomial ring* with coefficients R_0 and variables all U_i–symbols ($i = 1, \ldots, k$). In the following paragraphs, we associate with any local move type T a set of elements of P. These elements generate an ideal I_T so that the coset $\|X\|_\phi + I_T \in P/I_T$ does not change when applying a move of type T to X.

Let $T = (A_1, A_2, \Gamma_1, \Gamma_2, \phi)$ be a local move type. Let Φ_o be a choice of an element of C_i for each outer U_i-stratum of A_1. Since each outer stratum of A_1 uniquely corresponds to an outer stratum of A_2 of the same local type, we can at the same time interpret Φ_o as a choice of an element of C_i for each outer U_i-stratum of A_2. We denote the set of all states of $A_1 \setminus \Gamma_1, A_2 \setminus \Gamma_2$ that restrict to Φ_o on the outer strata by $\Sigma(A_1, \Phi_0), \Sigma(A_2, \Phi_0)$.

Theorem 3.29 *Let*

$$\gamma_{\Phi_o} = \sum_{\Phi \in \Sigma(A_1, \Phi_o)} |A_1 \setminus \Gamma_1|_\Phi - \sum_{\Phi \in \Sigma(A_2, \Phi_o)} |A_2 \setminus \Gamma_2|_\Phi.$$

Running over all choices of Φ_o, the γ_{Φ_o} generate an ideal $I_T \subset P$. The construction of I_T only depends on T and choices of $C_1, \ldots C_k$.

If X_1, X_2 are locally built upon $U_1, \ldots, U_k$ so that X_2 is obtained from X_1 by a move of type T, then $\|X_1\|_\phi - \|X_2\|_\phi \in I_T$.

Proof: The move relating X_1 with X_2 is defined by embeddings $\iota_1 \colon A_1 \to X_1$ and $\iota_2 \colon A_2 \to X_2$. For $i = 1, 2$, a stratum of X_i is either disjoint from $\iota_i(A_i)$ (first kind), or contain the ι_i-image of an outer stratum of A_i (second kind), or is the ι_i-image of an inner stratum of A_i (third kind). Let $\Theta_1(X_i), \Theta_2(X_i), \Theta_3(X_i)$ be the set of strata of X_i of the first, second respectively third kind. A choice Φ_o of a colour for each stratum of the first or second kind of X_1 can at the same time be interpreted as a choice of colours for the respective strata of X_2. Moreover, Φ_o gives rise to a choice of colours of outer strata of A_1, A_2 as well, that we keep denoting by Φ_o. In particular,

- if $s_1 \in \Theta_1(X_1)$ and s_2 is the corresponding stratum in $\Theta_1(X_2)$, then $|(s_1)_{\Phi_o}| = |(s_2)_{\Phi_o}|$ and $\epsilon_{\Phi_o(s_1)}(\delta(s_1)) = \epsilon_{\Phi_o(s_2)}(\delta(s_2))$.
- if $s_1 \in \Theta_2(X_1)$ and s_2 is the corresponding stratum in $\Theta_2(X_2)$, then $|(s_1)_{\Phi_o}| = |(s_2)_{\Phi_o}|$. If s is the outer stratum of A_2 that is mapped into s_2 by ι_2, then $\epsilon_{\Phi_o(s_2)}(\delta(s_2)) = \epsilon_{\Phi_o(s_2)}(\delta(s_1) \cdot \delta(s)) = \epsilon_{\Phi_o(s_1)}(\delta(s_1)) \cdot \epsilon_{\Phi_o(s)}(\delta(s))$.

Thus, we can obtain $\|X_1\|_\phi - \|X_2\|_\phi$ by summation of

$$\sum_{\Phi \in \Sigma(X_1, \Phi_o)} \prod_{s \in \Theta(X_1)} |s_\Phi| \cdot \epsilon_{\Phi(s)}(\delta(s))$$
$$- \sum_{\Phi \in \Sigma(X_2, \Phi_o)} \prod_{s \in \Theta(X_2)} |s_\Phi| \cdot \epsilon_{\Phi(s)}(\delta(s))$$
$$= \prod_{s \in \Theta_1(X_1)} |s_{\Phi_o}| \cdot \epsilon_{\Phi_o(s)}(\delta(s)) \cot$$
$$\left(\sum_{\Phi \in \Sigma(X_1, \Phi_o)} \prod_{s \in \Theta_2(X_1) \cup \Theta_3(X_1)} |s_\Phi| \cdot \epsilon_{\Phi(s)}(\delta(s)) \right.$$
$$\left. - \sum_{\Phi \in \Sigma(X_2, \Phi_o)} \prod_{s \in \Theta_2(X_2) \cup \Theta_3(X_2)} |s_\Phi| \cdot \epsilon_{\Phi(s)}(\delta(s)) \right)$$
$$= \gamma_{\Phi_o} \cdot \prod_{s \in \Theta_1(X_1)} |s_{\Phi_o}| \cdot \epsilon_{\Phi_o(s)}(\delta(s)) \cdot \prod_{s \in \Theta_2(X_1)} \epsilon_{\Phi_o(s)}(\delta(s))$$

over all choices of Φ_o, which apparently is a sum of elements of I_T. □

Remark 3.30 *Usage of ideal state sum invariants requires to test whether two elements of P yield the same coset modulo I_T. That problem can computationally be solved with* Gröbner *bases, see for example [KrRo00]: When choosing a so-called admissible monomial ordering on P, one can compute a Gröbner basis G of I_T, and then G can be used to compute the* normal form *of $p + I_T$ for any $p \in P$: Two elements $p, q \in P$ have the same normal form if and only if $p + I_T = q + I_T$.*

However, computing the Gröbner basis of an ideal can be very challenging, as the computational complexity is (in general) double exponential in the number and degree of its generators. In practical computations, the complexity depends on the choice of the monomial ordering of P. Below, we will introduce some simplifying assumptions, that may result in invariants that are weaker but easier to compute.

Remark 3.31 (and Definition) *When we consider a finite calculus with local move types $T_1, \ldots, T_l$, and X is built upon $X_1, \ldots, X_k$, then the coset $\mathcal{I}_\phi(X) := \|X\|_\phi + (I_{T_1} + \cdots I_{T_l})$ only depends on the finite calculus' connected component of X. This is what we call an* ideal state sum invariant *of X with respect to the finite calculus.*

3.6 Ideal state sum invariants of the Matveev-Piergallini calculus

Here, we focus on the finite calculus on simple polyhedra that is given by the T move. Invariants of that calculus give rise to homeomorphism invariants of closed 3-manifolds. In this section, we denote a Matveev-Piergallini move relating X_1 with X_2 by a local move type $T = (A_1, A_2, \Gamma_1, \Gamma_2, \phi)$ with embeddings $\iota_i \colon A_i \hookrightarrow X_i$ ($i = 1, 2$). For simplicity, we speak here of vertices, edges and 2-strata, rather than of U_3-, U_2- and U_1-strata of a simple polyhedron.

Example 3.32 *The construction of a state sum invariant starts with associating a finite set C_i to each local type U_i. Here, let $C_2 = C_3 = \{0\}$. The Matveev-Piergallini move, with 2-strata labelled as in the picture*

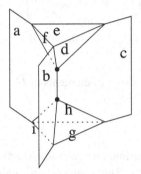 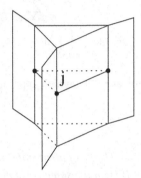

(where A_1 is on the left hand side of the picture, h is the outer stratum at the backside of A_1, j is the triangular stratum in the centre of A_2, and the outer strata of A_2 are labelled analogously to those of A_1) gives rise to the ideal $I_T \subset P$ generated by

$$\gamma_{a,\ldots,i} = (|a\,b\,c| \begin{vmatrix} a & b & c \\ d & e & f \end{vmatrix} \begin{vmatrix} a & b & c \\ g & h & i \end{vmatrix}$$
$$- \sum_{j \in C_1} |j| |f\,i\,j| |d\,g\,j| |e\,h\,j| \begin{vmatrix} a & f & e \\ j & h & i \end{vmatrix} \begin{vmatrix} b & d & f \\ j & i & g \end{vmatrix} \begin{vmatrix} c & e & d \\ j & g & h \end{vmatrix}) \cdot$$
$$|a||b||c||d||e||f||g||h||i|\,|a\,f\,e|\,|b\,d\,f|\,|c\,d\,e|\,|a\,i\,h|\,|b\,i\,g|\,|c\,g\,h|$$

for $a, b, c, d, e, f, g, h, i \in C_1$, where $|x|$ denotes the U_1-symbol associated with a 2-stratum of colour x.

Simplifying assumptions motivated by quantum invariants: Any quantum group at a $2n$-th root of unity gives rise to state sum invariants for the Matveev-Piergallini calculus taking values in $R_0 = \mathbb{C}$, called a *quantum 6j-invariant* or a *Turaev–Viro invariant* (see [Tur94]). The construction yields explicit values in $\mathbb{C}$ for each U_i-symbol ($i = 1, 2, 3$), and the sets C_1, C_2, C_3 of colours are

Aspects of TQFT and Computational Algebra 69

determined by the quantum group as well. Hence, one obtains the quantum 6j-invariant by evaluation of the ideal state sum invariant given by the same sets of colours.

The 6j-symbol evaluations given by a quantum group will generally provide properties that are not a consequence of the polynomial relations encoded in I_T. Here, we exploit such additional properties to obtain simplifications of the state sum formalism, yielding easier to compute ideal state sum invariants that may be weaker than the general ideal state sum invariants, but are non-trivial, as they still evaluate to a quantum 6j-invariant.

Discarding 3j-symbols: Each vertex v of a simple polyhedron is adjacent to four edges e_1, e_2, e_3, e_4, and for each edge there are exactly two vertices that the edge is adjacent with. Hence, if X is a simple polyhedron with a colouring Φ, the 3j-symbols in $|X|_\Phi$ are determined by the 6j-symbols in $|X|_\Phi$.

Replacing $\begin{vmatrix} a & b & c \\ d & e & f \end{vmatrix}$ with $\begin{vmatrix} a & b & c \\ d & e & f \end{vmatrix} \sqrt{|a\ b\ c|} \sqrt{|a\ e\ f|} \sqrt{|b\ d\ f|} \sqrt{|c\ d\ e|}$ discards all 3j-symbols in the formalism, without changing $|X|_\Phi$. Thus, the construction of an ideal state sum invariant for the Matveev-Piergallini calculus in Example 3.32 can be simplified by dropping any reference to the 3j-symbols.

Admissible colourings: Some 3j- and 6j-symbols evaluate to zero or to one, in the quantum invariants associated with a specific quantum group. For example, in the case of the quantum group $U_q(\mathfrak{sl}_2)$, we have $C_1 = \{0, \ldots, n-1\}$, $|0| = |0\ 0\ 0| = \begin{vmatrix} 0 & 0 & 0 \\ 0 & 0 & 0 \end{vmatrix} = 1$ and we have $|a\ b\ c| \neq 0$ *only* if $a, b, c \in C_1$ satisfy the condition that $a + b + c$ is even, $a + b \geq c$, $a + c \geq b$ and $b + c \geq a$. In that case, (a, b, c) is called *admissible*.

Hence, setting $|0| = |0\ 0\ 0| = \begin{vmatrix} 0 & 0 & 0 \\ 0 & 0 & 0 \end{vmatrix} = 1$ and setting $|a\ b\ c| = 0$ whenever (a, b, c) is not admissible, and also setting $\begin{vmatrix} a & b & c \\ d & e & f \end{vmatrix} = 0$ whenever at least one of $(a, b, c), (a, e, f), (b, d, f), (c, d, e)$ is not admissible, results in a simplified ideal state sum invariant for closed 3-manifolds that is still at least as strong as the invariant obtained from $U_q(\mathfrak{sl}_2)$.

For the ϵ-invariant [MaSo96] one has $C_1 = \{1, 2\}$, and $|a\ b\ c| = 0$ whenever $a + b + c = 1$. This gives rise to another notion of admissible colourings.

3.6.1 Experimental results

King [Kin07a] showed by practical experiments that the simplified ideal state sum invariant distinguishes more homeomorphism types of closed 3-manifolds than the corresponding quantum 6j-invariants (i.e., invariants with the same set C_1 and the same admissibility conditions). The experiment was conducted

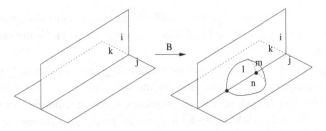

Figure 3.17 The bubble move

on all 1900 closed orientable irreducible 3-manifolds of Matveeev complexity ≤ 9. We studied ideal state sum invariants tv_2 and tv_3^+ with 2 respectively with 3 colours for U_1-strata, with 6j-symbols subject to implying assumptions as explained above. For tv_3^+, we used a non-trivial involution in order to take into account local orientations of U_1-strata.

Among the 1900 samples, tv_2 alone distinguishes 134 homeomorphism types and tv_3^+ distinguishes 242 homeomorphism types. In both cases, this is considerably better than quantum 6j invariants with 2 respectively with 3 colours. However, homology seems stronger, as it can distinguish 272 homeomorphism types. But in a way, ideal state sum invariants complement homology: A combination of both can distinguish 764 homeomorphism types.

3.6.2 Non-modular TQFT

For the reasons given in Section 3.3.4, for a study of the Andrews–Curtis conjecture one should consider non-modular TQFT. The TQFTs constructed in [Tur94] are all modular. We outline in this subsection why the ideal invariants constructed above are not modular.

Here is one way to look at the (AC')-conjecture: Let us first consider the case, where K and L are two special 2-complexes with trivial fundamental group and the same Euler characteristic. A homotopy equivalence $f : K \to L$ is a simple homotopy equivalence if and only if the Whitehead torsion $\tau(f) = 0$; $\tau(f)$ is an element of the Whitehead group $Wh(\pi_1(L))$, which is 0 for trivial $\pi_1(L)$, see [I]. So, K and L are simple homotopy equivalent and are related by the local moves T_1, T_2, T_3, the *bubble move* (see Figure 3.17) and their inverses. As in the case of trivial fundamental group, the crucial question for (AC') is whether it is possible to relate K and L by local moves without using any bubble move.

Let $\mathcal{Z}$ be a TQFT of special 2-complexes that is invariant under all moves of

Aspects of TQFT and Computational Algebra

the local calculus except for the bubble move. Let K_1 be a special 2-complex and K_2 be obtained from K_1 by a bubble move. Applying the bubble move is essentially the same as a one-point union with a 2-sphere. Hence, if $\mathcal{Z}$ is modular and if $z := \mathcal{Z}(S^2)$, then $\mathcal{Z}(K_2) = z \cdot \mathcal{Z}(K_1)$.

If K, L are simple homotopy equivalent and have the same Euler characteristic, the number of bubble moves is equal to the number of inverse bubble moves in any sequence of local moves relating K and L. Thus, if z is invertible, then $\mathcal{Z}(K) = \mathcal{Z}(L)$. One could try to pass to a ring quotient in which z is sent to zero and is thus not invertible. However, by [BoQu05], this would still not allow to distinguish different 3-deformation types.

But the situation with the ideal invariants is different. Again, let K_1 and K_2 be related by a bubble move as in the picture. Examples in [Kin07b] show that in general, the ideal state sum invariant of K_2 is not simply the product of the ideal state sum invariants of K_1 and of the 2-sphere, which would be the case for a modular invariant. The underlying reason is as follows: Let Φ_o be a colouring for the outer strata of the picture (i.e., a choice of $i, j, k \in C_1$), and let Φ_1, Φ_2 be colourings of K_1, K_2 extending Φ_o and coinciding outside of the local picture. It is not difficult to verify that for each Φ_o there is a constant z_{Φ_o} such that $|K_2|_{\Phi_2} = z_{\Phi_o}|K_1|_{\Phi_1}$ for any choice of Φ_1, Φ_2: Indeed, $z_{\Phi_o} = \sum_{l,m,n \in C_1} |l| \cdot |m| \cdot |n| \cdot \begin{vmatrix} i & j & k \\ n & l & m \end{vmatrix} \cdot \begin{vmatrix} j & i & k \\ l & n & m \end{vmatrix}$. Apparently the constant depends on i, j, k, that's to say, it depends on Φ_o. Thus, one obtains $\|K_2\|_\phi$ by multiplying the different summands of $\|K_1\|_\phi$ with *different* factors. By the results exposed in Section 3.3.4, in a modular state sum invariant, the factors modulo the relation ideal were the same for all summands, namely $\|S^2\|_\phi$.

Since a bubble move does not result in the multiplication of $\mathcal{I}_\phi(\cdot)$ with a constant factor, it is not a modular invariant. Thus, the negative result from [BoQu05] does not apply, and thus one might hope that the ideal state sum invariants may be able to distinguish two simple homotopy equivalent special 2-complexes with the same Euler characteristic. However, the experiments described by King in [Kin07b] did not result in the detection of a counterexample to the (AC') conjecture. At this point, we can only speculate whether our nonmodular TQFT has a chance to detect (AC') counter examples at all.

We remark that, if an invariant is desired that is multiplicative with respect to connected sum, one can modify the construction of ideal state sum invariants by adding further generators to the relation ideal, see [Kin07b].

4

Labelled Oriented Trees and the Whitehead-Conjecture

Stephan Rosebrock

4.1 Introduction

In this chapter new developments concerning the Whitehead conjecture are presented. The Whitehead conjecture states that subcomplexes of aspherical 2-complexes are aspherical. Despite some effort in this area the Whitehead conjecture still remains open.

Labeled oriented trees play a crucial role in the study of the finite case of the conjecture. The related chapter in [I], Chapter X, only touches labelled oriented trees. Here we focus on them but also present other new results on the Whitehead conjecture.

In 2007 another survey on the Whitehead conjecture appeared (see ROSEBROCK [Ros07]).

This chapter is organized as follows: In Section 2 we present a result of LUFT that the existence of a finite counterexample implies the existence of an infinite counterexample to the Whitehead conjecture. Also labelled oriented trees are defined. In Section 3 we define spherical diagrams and Whiteheadgraphs. Several classes of aspherical labelled oriented trees are presented in Section 4. In Section 5 the asphericity of injective LOTs is shown. This is done essentially by a new test on relative asphericity. Section 6 presents a non-aspherical LOG with certain interesting properties. Implications of the asphericity of injective LOTs to virtual knots are presented in the short Section 7. In Section 8 written by Jens Harlander L^2-homology and its implications to the Whitehead conjecture are discussed.

BESTVINA and BRADY have shown 1997 that either the Whitehead conjecture or the Eilenberg-Ganea conjecture must be wrong (see [BeBr97]). There exists an alternative proof by HOWIE (see [How99]). Details are presented in Chapter 6 of this volume.

4.2 The finite and the infinite case

In [I], Chapter X Theorem 4.1 of Howie states that if there exists a counterexample to the Whitehead conjecture consisting of an aspherical 2-complex L containing a non-aspherical subcomplex K, then there must be such a pair (L, K) either of type (a) where L is finite, contractible, and obtained from K by attaching a single 2-cell or of type (b) where L is the union of an infinite chain of non-aspherical subcomplexes. Luft [Luf96] has shown 1996 that the existence of a counterexample of type (a) implies the existence of one of type (b):

Theorem 4.1 *If there exists a finite contractible counterexample to the Whitehead conjecture, then there exists a counterexample of type (b).*

Proof: Assume L_0 is a finite, contractible aspherical 2-complex and $K'_0 \subset L_0$ a non-aspherical subcomplex. Let (K'_i, L_i) with $i = 1, 2, 3, \ldots$ be infinitely many copies of (K'_0, L_0). Let $x_i \in K'_i$ be a point for each $i = 1, 2, 3, \ldots$. Denote $\mathbb{R}_+ = \{x \in \mathbb{R} \mid x \geq 0\}$. Let

$$L = \mathbb{R}_+ \cup \bigcup_{i=0}^{\infty} L_i$$

where x_i is identified with $i \in \mathbb{R}_+$ ($i = 0, 1, 2, \ldots$). Certainly, L is aspherical. Define

$$K_i = [0, i] \cup \bigcup_{k=0}^{i-1} L_k \cup K'_i \quad \text{for } i = 1, 2, 3, \ldots$$

Each K_i is non-aspherical and the inclusion map $K_i \to K_{i+1}$ is nullhomotopic for all $i = 1, 2, 3, \ldots$. So this is a counterexample of type (b). □

In the same paper Luft also presents a simpler proof of [I], Chapter X Theorem 4.1 than the original proof of Howie.

In [I], Chapter X Theorem 4.6 it is shown that (provided that the Andrews-Curtis conjecture is true) the existence of a counterexample of type (a) implies the existence of a non-aspherical labelled oriented tree. A *labelled oriented graph* (LOG) is an oriented graph $\mathcal{G}$ on vertices **x** and edges **e**, where each oriented edge is labelled by a vertex. Associated with it is the *LOG-presentation* $P(\mathcal{G}) = \langle \mathbf{x} \mid \{r_e\}_{e \in \mathbf{e}} \rangle$. If e is an edge that starts at x, ends at y and is labelled by z, then $r_e = xz(zy)^{-1}$. A *LOG-complex* $K(\mathcal{G})$ is the standard 2-complex associated with the LOG-presentation $P(\mathcal{G})$, and a *LOG-group* $G(\mathcal{G})$, is the group defined by the LOG-presentation. We say a labelled oriented graph is *aspherical* if its associated LOG-complex is aspherical. A *labelled oriented tree* (LOT) is a labelled oriented graph where the underlying graph is a tree. A non-aspherical labelled oriented tree would be a Whitehead counterexample since adding x_i as a relator for any generator x_i gives a balanced presentation of the trivial group which is aspherical. So labelled oriented trees are the ideal test complexes for the finite case of the Whitehead conjecture.

In [HaRo03] HARLANDER and ROSEBROCK construct a singular 3-manifold M (with exactly one non-manifold point which has a surface link) out of any given LOT $\mathcal{G}$, which is homotopy equivalent to $K(\mathcal{G})$. M is a 3-manifold if $\mathcal{G}$ is the Wirtinger-presentation of a knot L. M mimics the knot space $S^3 - L$ otherwise.

4.3 Spherical diagrams

Spherical diagrams are important for the study of asphericity.

Definition 4.2 *A spherical diagram over a 2-complex K given by a presentation $\langle \mathbf{x} \mid \mathbf{r} \rangle$ is a combinatorial map $f: C \to K$, where C is a cell decomposition of the 2-sphere and the restriction of f to each open cell of C is a homeomorphism onto its image. Moreover,*

- *oriented edges $e \in C^{(1)}$ are labelled by $f(e)$;*
- *the word read off the boundary path of an inner or the outer region of $C^{(1)}$ is a cyclic permutation of a word r^ϵ, where $\epsilon = \pm 1$ and $r \in \mathbf{r}$.*

If K is non-aspherical, then there exists a spherical diagram which realizes a nontrivial element of $\pi_2(K)$. In fact, $\pi_2(K)$ is generated by spherical diagrams (see [I], Chapter V Theorem 1.3, p. 162 and the literature cited there). So in order to check whether a 2-complex is aspherical or not it is enough to check spherical diagrams.

A spherical diagram $f: C \to K^2$ is *reducible*, if there exists a pair of 2-cells in C with a common edge e, such that both 2-cells are mapped to K by folding over e. A 2-complex K is called *diagrammatically reducible* (DR), if every spherical diagram over K is reducible. Certainly if K is DR then K is aspherical since any spherical diagram over a DR 2-complex can be reduced to the trivial diagram.

Definition 4.3 *Let Γ be a graph and Γ_0 be a subgraph (which could be empty).*

1 An edge cycle $c = e_1 \ldots e_q$ in Γ is called homology reduced *if it does not* contain a pair of edges e_i and e_j so that $e_j = \bar{e}_i$, where $\bar{e}_i$ is the edge e_i with opposite orientation.

2 An edge cycle $c = e_1 \ldots e_q$ is said to be homology reduced relative to Γ_0 if it does not contain a pair of edges e_i and e_j of $\Gamma - \Gamma_0$ so that $e_j = \bar{e}_i$.

If v is a vertex of a 2-complex K then the link $\mathrm{Lk}(K, v)$ is the boundary of a regular neighborhood of v in K equipped with the induced cell decomposition. So $\mathrm{Lk}(K, v)$ is a graph. Let the 2-complex K be the standard 2-complex associated with a group presentation $P = \langle \mathbf{x} \mid \mathbf{r} \rangle$. K has a single vertex v. So $\mathrm{Lk}(K, v)$ is the Whiteheadgraph of K which we also denote by $\mathrm{Lk}(K)$ or by $W(P)$. We refer to the edges in $\mathrm{Lk}(K)$ as *corners*, since they can be thought of as the corners of the 2-cells of K. The Whiteheadgraph $W(P)$ is a non-oriented graph on vertices $\{x^+, x^- \mid x \in \mathbf{x}\}$, where x^+ is a point of the oriented edge x of K close to the beginning of that edge, and x^- is a point close to the ending of that edge. Vertices x^ϵ and y^δ, $(x, y \in \mathbf{x},\ \epsilon, \delta \in \{\pm\})$, are connected by an edge in $\mathrm{Lk}(K)$ if there is a 2-cell in K with a corner connecting the two points. The *positive graph* $W^+(P) \subset W(P)$ is the *full subgraph* on the vertex set $\{x^+ \mid x \in \mathbf{x}\}$ (i.e. $W^+(P)$ has the vertex set $\{x^+ \mid x \in \mathbf{x}\}$ and all the edges of $W(P)$ which run between these vertices), the *negative graph* $W^-(P) \subset W(P)$ is the full subgraph on the vertex set $\{x^- \mid x \in \mathbf{x}\}$.

Let K_0 be a subcomplex of the 2-complex K. Then $\mathrm{Lk}(K_0)$ is a subgraph of $\mathrm{Lk}(K)$. Let $f \colon C \to K$ be a spherical diagram and v be a vertex of the 2-sphere C. The map f induces a combinatorial map $f_L \colon \mathrm{Lk}(C, v) \to \mathrm{Lk}(K)$. Note that $\mathrm{Lk}(C, v)$ is a circle and the image of that circle, oriented clockwise, is a *cycle of corners* $\alpha(v) = \alpha_1 \ldots \alpha_q$, that is a closed edge path, in $\mathrm{Lk}(K)$. We say that the diagram $f \colon C \to K$ is *vertex reduced at v relative to K_0* if the cycle $\alpha(v)$ in $\mathrm{Lk}(K)$ is homology reduced relative to $\mathrm{Lk}(K_0)$. We say that the diagram is *vertex reduced relative to K_0* if it is vertex reduced relative to K_0 at all its vertices. The 2-complex K is called *vertex aspherical relative to K_0*, VA relative to K_0 for short, if for every spherical diagram $f \colon C \to K$ that is vertex reduced relative to K_0 we have $f(C) \subseteq K_0$.

If we omit "relative to" we implicitly imply relative to the empty set $\emptyset$, even if a subcomplex is present. For example if we say a spherical diagram is vertex reduced we mean vertex reduced relative to $\emptyset$. Consequently a 2-complex K is called *vertex aspherical* (VA) if there is no vertex reduced spherical diagram over K. For a 2-complex we have DR $\Rightarrow$ VA $\Rightarrow$ aspherical.

Theorem 4.4 [HaRo17] *If K is VA relative to K_0, then $\pi_2(K)$ is generated, as $\pi_1(K)$-module, by the image of $\pi_2(K_0)$ under the map induced by inclusion. In particular, if K_0 is aspherical, then so is K.*

Proof: The statement follows since $\pi_2(K)$ is generated by spherical diagrams and every spherical diagram $f : C \to K$ is homotopic to a spherical diagram $f' : C' \to K$ that is vertex reduced. □

Let $\mathcal{G}$ be a LOG and $K(\mathcal{G})$ the corresponding LOG-complex. Given a spherical diagram $f : C \to K(\mathcal{G})$ we can draw its dual by replacing the square 2-cells by crossings. This idea is originally from Wolfgang Metzler and already described in [I], Chapter X. We undercross when labels change. The process is depicted for a single 2-cell of C in Figure 4.1. [ht] We define an orientation

Figure 4.1 Dualizing a spherical diagram

of the link by requiring that when traveling along the link we encounter the 1-cells as pointing to the left and label the components between undercrossing and undercrossing by the labels of the original 1-cells of C. This leads to an oriented labelled link projection L on the 2-sphere C which contains all the information of the diagram.

Observe that L lifts to $\mathcal{G}$. More precise, there is an induced map $f_* : L \to \mathcal{G}$ defined by mapping an arc between undercrossing and undercrossing of L labelled a to the vertex $a \in \mathcal{G}$. When we reach a crossing in L we pass the corresponding edge in $\mathcal{G}$. If $\mathcal{G}$ is a LOT then $f_*(L)$ is contractable since $\mathcal{G}$ is a tree.

There is also an induced homomorphism of the link group to the LOG-group $f^* : \pi_1(S^3 - L) \to G(\mathcal{G})$ defined by mapping the dual generator of an arc of L between undercrossing and undercrossing labelled a to the generator $a \in G(\mathcal{G})$. If f is surjective then f^* is surjective which implies that a trivial knot L may only map to a LOT presenting the infinite cyclic group.

4.4 Classes of aspherical LOTs

Wirtinger presentations of knot groups are LOT presentations. They are aspherical by a result of PAPAKYRIKOPOULOS (see [I], [Pa57]). Many LOTs are not Wirtinger presentations of knots (see [Ros94]). So, there is still a lot of work to do. In [I], [Ho85] it is shown that LOTs of diameter 3 are aspherical.

The following Theorem is due to GERSTEN (see [I], [Ger87$_1$]) stated in a more general context:

Theorem 4.5 *If the negative graph or the positive graph of a LOT $\mathcal{G}$ is a tree, then $K(\mathcal{G})$ is DR.*

Proof: We can define a map $g: K(\mathcal{G}) \to S$, where S is the unit circle with a single vertex u and a single edge e oriented clockwise, by mapping the vertex of $K(\mathcal{G})$ to u, and an oriented edge of $K(\mathcal{G})$ in an orientation preserving manner to e. Given a spherical diagram $f: C \to K(\mathcal{G})$ we can continuously extend the map $h = g \circ f$ to the 2-skeleton of C since relators of LOTs have exponent sum 0. We can now lift h to $\tilde{h}: C \to \tilde{S}$, where $\tilde{S}$ is the universal covering of S. Note that $\tilde{S}$ is the real number line $\mathbb{R}$ with its vertices located at the integers. The map $\tilde{h}$ takes on a maximum at some vertex v and a minimum at some vertex v'. Lk$(K(\mathcal{G}), v)$ maps to the negative graph and Lk$(K(\mathcal{G}), v')$ maps to the positive graph. One of those is a tree which implies that its image is contractible. So f is DR. □

Theorem 4.5 was generalized by J. BARMAK and E. MINIAN (see [BaMi16]). They invented the *I-test* for a 2-complex which implies DR. The conditions required by the I-test imply that the image of a spherical diagram lifted to the universal cover always collapses to a 1-complex which in turn is equivalent to being DR. An alternative proof, where the I-matrix is viewed as a height matrix, was worked out by J. HARLANDER and the author (unpublished). This proof is presented here:

Let $P = \langle x_1, ..., x_n \mid r_1, ..., r_m \rangle$ be a presentation, $n \geq m$, K be the standard 2-complex constructed from P and $\tilde{K}$ be its universal covering. Note that the 1-skeleton of $\tilde{K}$ is the Cayley graph. Let $\tilde{x}_i$ be the edge in $\tilde{K}$ that starts at 1 and maps to x_i in K. Let $\tilde{r}_j \in \tilde{K}$ be the 2-cell whose boundary reads the word r_j in the Cayley graph that starts at 1. Let G be the group defined by P. Assume there is an epimorphism $\phi: G \to \mathbb{Z}$. Any such map yields a height function $\tilde{K} \to \mathbb{R}$, sending a vertex g to $\bar{g} = \phi(g)$. The height of a cell in $\tilde{K}$ is defined to be the height of its base point. So the height of the edge $g\tilde{x}_i$ is $\bar{g}$ and the height of the 2-cell $g\tilde{r}_j$ is $\bar{g}$.

Let M be the matrix with entry $M_{i,j}$ being the family of heights of the edges labeled x_i in $\tilde{r}_j$. We call M the *height matrix* of P.

A matrix M with entries consisting of finite families of numbers is called *good* if

(1) the entry $M_{k,k}$ for $1 \leq k \leq m$ is non-empty,
(2) the maximum λ of $M_{k,k}$ is the maximum of the entire k-th row $\bigcup_{j=1}^{n} M_{k,j}$,
(3) the multiplicity of λ in the partial row $\bigcup_{j=k}^{m} M_{k,j}$ is one.

The I-test now checks whether the height matrix of a presentation is good:

Theorem 4.6 *Assume the height matrix M is good. Then K is DR.*

Proof: Let S be a spherical diagram over K. We lift it and obtain a spherical diagram $\tilde{S}$ over $\tilde{K}$. Let α be the maximal 2-cell height that occurs in $\tilde{S}$. Let

$$t = min\{1 \leq j \leq m \mid g\tilde{r}_j \text{ is a 2-cell of maximal height } \alpha \text{ in } \tilde{S}\}.$$

Choose a 2-cell $h\tilde{r}_t$ where $\bar{h} = \alpha$. In that 2-cell choose an edge e with edge label x_t of maximal height $\lambda \in M_{t,t}$. This is possible because of conditions (1) and (2) in the definition of "good". Note that $e = g_0 \tilde{x}_t$ for some $g_0 \in G$. Let

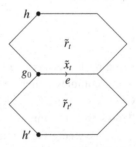

Figure 4.2 Part of a spherical diagram $\tilde{S}$

$h'\tilde{r}_{t'}$ be the other 2-cell in $\tilde{S}$ that contains e. It follows that $\bar{h}' \leq \bar{h}$ since $h\tilde{r}_t$ is a 2-cell of maximal height. Furthermore, we have

$$\bar{g}_0 = \bar{h} + \lambda = \bar{h}' + m_{t,t'},$$

where $m_{t,t'} \in M_{t,t'}$. If $\bar{h}' < \bar{h}$ then $m_{t,t'} > \lambda$, which contradicts condition (2). Thus $\bar{h}' = \bar{h}$ and it follows that $t' \geq t$ by the choice of t. Since $\bar{h}' = \bar{h}$ we also have $\lambda = m_{t,t'}$. Condition (3) now implies that $t = t'$ and that the edge e in $\tilde{S}$ is

a folding edge. □

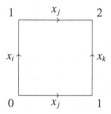

Figure 4.3 Heights of a LOT-relation $x_i x_j = x_j x_k$.

In the situation of LOGs we always have an epimorphism of the LOG-group onto $\mathbb{Z}$ by mapping each generator of the corresponding LOG-presentation to 1. In Figure 4.3 the heights of a LOT-relation $x_i x_j = x_j x_k$ are shown by putting the basepoint at the vertex labelled 0.

Barmak and Minian show, that a LOG where the left or right graph is a tree satisfies the I-test. They also present an example of a LOT with cycles in the left and the right graph which satisfies the I-test:

Example 4.7 *The LOT of Figure 4.4 satisfies the I-test. The corresponding height matrix is*

$$\begin{pmatrix} 0,1 & & & & & 0 \\ 0 & 0 & & & & \\ & & 1 & 1 & & \\ & & 0,1 & 0 & & 0,1 \\ 1 & & & 0,1 & 1 & \end{pmatrix}$$

which is easily seen to be good.

Figure 4.4 A LOT satisfying the I-test.

LOTs are DR if they satisfy the small cancellation conditions C(4), T(4) (see [I], [LySc77]), or more generally the weight test (see [I], Chapter X Theorem 2.2) or the cycle test (see HUCK/ROSEBROCK [I], [HuRo92]).

In [Ros10] the author introduces a notion of complexity: Given a LOT $\mathcal{G}$

with LOT presentation $P = \langle \mathbf{x} \mid \mathbf{r} \rangle$, we say that P has *complexity n*, provided that there exists a subset $\mathbf{s} = \{x_{i_1}, \ldots, x_{i_n}\}$ of the set of generators consisting of n elements, such that the following process defines every generator of P to be good and there is no such set consisting of $n - 1$ elements:

1 The elements of $\mathbf{s}$ are good.
2 If $xy = yz$ or $zy = yx$ is a relator of P and x, y are good, then so is z.

We say that P is *derived by* $\mathbf{s}$.

Theorem 4.8 *LOTs of complexity 2 are aspherical.*

Proof: A LOT of complexity n can be transformed into a presentation consisting of n generators and $n - 1$ relators: The generators of $\mathbf{x}$ not in $\mathbf{s}$ may be eliminated by Q^{**} transformations together with relators from the LOT.

If a LOT has complexity 2, then this process leads to a 1-relator presentation which is known to be aspherical by a result of Lyndon (see [I], [Ly50]) if the relator is not a proper power. Since LOTs abelianize to the infinite cyclic group the relator is not a proper power. □

If the number of different edge labels of a LOT is n, then its complexity is $\leq n$. This is because any LOT may be derived by its edge labels. So we have as a consequence:

Corollary 4.9 *A LOT with at most two different edge labels is aspherical.*

A *sub-LOG* of a labelled oriented graph $\mathcal{G}$ is a connected subgraph $\mathcal{H}$ (containing at least one edge) such that every edge label of $\mathcal{H}$ is a vertex of $\mathcal{H}$. A sub-LOG $\mathcal{H}$ is *proper* if it is not all of $\mathcal{G}$. A labelled oriented graph is called *compressed* if no edge is labelled with one of its vertices. It is called *boundary reducible* if there is a boundary vertex (that is a vertex incident to exactly one edge) that does not occur as edge label, and *boundary reduced* otherwise. A labelled oriented graph is called *interior reducible* if there is a vertex with two adjacent edges with the same label that either point away or towards that vertex, and *interior reduced* otherwise. A labelled oriented graph which is boundary reduced, interior reduced and compressed is called *reduced*. Every LOT can be transformed into a reduced LOT by a simple homotopy.

MORITZ CHRISTMANN and TIMO DE WOLFF [ChdW14] show that an interior reduced LOT $\mathcal{G}$ with m vertices has complexity at most $(m + 1)/2$. The proof is constructive, they explicitly construct a set $\mathbf{s} \in \mathbf{x}$ which derives $\mathbf{x}$. They build the set $\mathbf{s}$ successively $s_1 \subset s_2 \subset \ldots \subset s_k = \mathbf{s}$ where s_1 consists of an arbitrary generator. To build s_k from s_{k-1} they add a generator which is an edge label of an edge where one boundary vertex x_i may be derived from s_{k-1} and the other one x_j not. So they derive x_j for free.

In the same paper it is shown that there exist LOTs of maximal complexity $(m+1)/2$ with m vertices. They explicitly describe all these LOTs. They are all composed of sub-LOTs given in Figure 4.5 (with any orientation of its edges) by identifying vertices. As mentioned by Christmann and de Wolff such

Figure 4.5 A building block of a LOT of maximal complexity

a LOT of maximal complexity is aspherical. This is because on the fundamental group level we have an amalgamated product which implies that the inclusion induced homomorphisms are injective. The components given in Figure 4.5 are aspherical. The result uses a Theorem of Whitehead (see Gersten [I], [Ge87$_2$], Theorem 5.1).

Another class of DR LOTs was worked out by the author in [Ros00] based on work of KLYACHKO [Kly93]. We need Klyachkos result in the following version: Let C be a cell decomposition of an oriented 2-sphere without vertices of valency one. On the boundary of each 2-cell runs a car without stops in the direction of the orientation of C. The cars are not allowed to go below some speed $\epsilon > 0$. Then there are at least two points of total collision in the 1–skeleton of C. *Total collision* means: If a point Q in the 1–skeleton of C is in the boundary of exactly n 2–cells, then Q is a point of total collision, if for some time all n cars are in Q.

So a 2-complex K can be shown to be DR in the following way: Assume that for any given spherical diagram $f \colon C \to K$ we can define motions of cars on the boundaries of the 2-cells of C such that there cannot be a point of total collision if f is reduced. Then K is DR. This is possible for several classes of LOTs. We present one class here:

Theorem 4.10 *The standard 2-complex $K(P)$ modeled on the presentation*

$$P = \langle x_1, \ldots, x_n \mid x_i w = w x_{i+1}, 1 \leq i < n \rangle$$

is DR for every $n \geq 1$ and every nonempty word w in the free group on $x_1, \ldots, x_n$ if the relators in P are cyclically reduced.

The presentations mentioned in this theorem are examples of word labelled oriented trees. A *word labelled oriented graph* (WLOG) is an oriented graph

$\mathcal{G}$ on vertices $X = \{x_1, \ldots, x_n\}$, where each oriented edge is labelled by a nonempty word in $X^{\pm 1}$. A *word labelled oriented tree* (WLOT) is a WLOG such that the underlying graph $\mathcal{G}$ is a tree. Associated with a word labelled oriented graph comes a *WLOG–presentation* $P(\mathcal{G})$ on generators X and relators in one-to-one correspondence with edges in $\mathcal{G}$. For an edge with initial vertex x_i, terminal vertex x_j, labelled w, we have a relation $x_i w = w x_j$. A WLOG may be easily transferred into a LOG by Q^{**}-operations. Thereby a WLOT is transferred into a LOT. For example the WLOG relator $x(ab^{-1}) = (ab^{-1})y$ can be transformed into $xa = az$ and $yb = bz$ with new generator z.

Proof: (of Theorem 4.10) Let $f: C \to K(P)$ be a reduced spherical diagram. We will define motions of cars on the 2-cells of C such that there are no points of total collision.

Let $x_i w = w x_{i+1}$ be any relator of P. In Fig. 4.6 the relator and its inverse

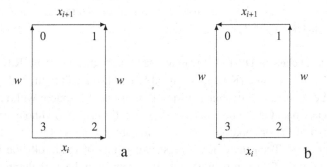

Figure 4.6 Car around a relator and its inverse

are shown. The numbers at the vertices describe at what time the car is where in any occurrence of the relator (Fig. 4.6 a) or its inverse (Fig. 4.6 b) in C. The times are given modulo 4. Thereby the edge labelled w may consist of several edges in the diagram. Assume that the car has constant velocity between times k and $k + 1$ for $0 \le k \le 4$.

P is such that for any generator x_j there is at most one 2-cell of $K(P)$ which has an edge labelled x_j which is in the image of edges of C with time 0 to 1. So in C there is no total collision at the times $0 < t < 1$ or by the same arguments $2 < t < 3$. Assume there would be a total collision at the time 2 at a vertex $Q \in C$. Since the relators of P are cyclically reduced, the valence of Q (i.e. the number of edges incident to Q) is bigger than one. If D is a preimage of the relator $R : x_i w = w x_{i+1}$ with Q in its boundary, then x_i is the label of a 1-cell with Q in its boundary, bounding D and another 2-cell $E \ne D$. But the only

2-cell having x_i at the time 2 in its boundary is R. If E was mapped to R, then the time at Q in E would be 3. The same argument works for all integer times.

So, in order to show Theorem 4.10 it remains to check the time $1 < t < 2$ (or $3 < t < 4 = 0$): A total collision in the middle of some edge of w cannot happen because of different orientations of the corresponding edges. A total collision at a vertex $Q \in C$ occurring at the time $1 < t < 2$ cannot happen if Q has valence 1 since w is a reduced word. But for a higher valence of Q, the labels do not fit. The same argument works for $3 < t < 4 = 0$. □

HARLANDER and ROSEBROCK have constructed in [HaRo15] a class of DR WLOGs by assigning weights to (potential) preimages in reduced spherical diagrams and thereby achieving a contradiction to the Euler-Characteristic of the 2-sphere. Here is the result without proof:

Theorem 4.11 *Let $P(\mathcal{G}) = \langle x_1, \ldots, x_n \mid r_1, \ldots, r_k \rangle$ be a WLOG-presentation coming from a word labelled oriented graph $\mathcal{G}$. Then each relation r_i is of the form $x_{\alpha(i)} w_i = w_i x_{\beta(i)}$ where w_i is a word $w_i = t_{i,1} \ldots t_{i,s_i}$ with $t_{i,j} \in \{x_1, \ldots, x_n\}^{\pm 1}$. We assume $s_i \geq 2$ for $i = 1, \ldots, k$. We further assume that*

1 *Each relator is cyclically reduced.*
2 *For each relator r_i the words $x_{\alpha(i)} t_{i,1}$, $t_{i,1}^{-1} x_{\alpha(i)}$, $x_{\beta(i)} t_{i,s_i}^{-1}$, $t_{i,s_i} x_{\beta(i)}$ are not pieces (i.e. these words and their inverses are not subwords in another relator or in the same relator at another place).*
3 *No word $w_i^{\pm 1}$ is a subword of some w_j for $j \neq i$,*
4 *No word w_i has the form $x_j^{\pm m}$ for $m \geq 2$.*

Then $K(\mathcal{G})$ is DR.

Another class of aspherical LOTs can be achieved by a lemma of STALLINGS in [Sta87]. It has been applied by Stallings and others to show that certain equations over groups are solvable and that certain 2-complexes are aspherical (see for instance GERSTEN [I], [Ge87$_1$]):

Lemma 4.12 *Every finite, nonempty, directed graph on the 2-sphere without isolated vertices has at least two consistent items.*

Here, a *consistent item* is either a *source* (i.e. a vertex from which all adjacent edges point away), a *sink* (i.e. a vertex towards which all adjacent edges point) or a *consistently oriented region* (i.e. a simply connected region whose boundary consists of edges, oriented all clockwise or all counterclockwise).

HUCK and ROSEBROCK have used this lemma in [HuRo07] to construct a class of aspherical LOTs by reversing the orientation of edges in (potential) reduced spherical diagrams. We demonstrate the idea of reversing edges in a simple example:

Example 4.13 *The 2-complex $K(P)$ modeled on the LOT-presentation*

$$P = \langle a, b, c, d, e, f \mid ac = cb, bd = dc, db = bc, df = fe, ea = af \rangle$$

is DR.

Proof: Assume there is a reduced spherical diagram C over $K(P)$. As for all LOG-complexes, the 2-cells of $K(P)$, and hence the 2-cells of C, are not consistently oriented since each relation reads two positive and two negative generators. If $W^+(P)$ and $W^-(P)$ were forests, no sinks or sources could occur in C, directly contradicting Stalling's Lemma; but, this is not the case. In Fig. 4.7, depicting $W(P)$, the vertices on the top row are $a^+, b^+, \ldots, f^+$ and on the

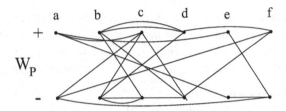

Figure 4.7 The Whiteheadgraph of P

lower row are $a^-, b^-, \ldots, f^-$. It shows that $W^+(P)$ and $W^-(P)$ each contain (up to proper powers) a single reduced cycle. These cycles are $d^+ \longrightarrow b^+ \longrightarrow d^+$ in $W^+(P)$ and $b^- \longrightarrow c^- \longrightarrow b^-$ in $W^-(P)$. The sinks and sources, corresponding to these cycles, that may occur in C are depicted in Fig. 4.8 (of course, proper powers of these cycles could occur also).

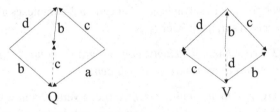

Figure 4.8 sink and source in C

We destroy such sinks or sources in C by reversing edges as follows. In every sink that occurs in C we reverse the dashed edge labelled c and, if the sink-cycle is a proper power, we reverse all edges labelled c at the sink. Analogously, we reverse the dashed edges labelled d in every source of C. To prove (by Stalling's Lemma) that $K(P)$ is DR, we need to show that no new sinks or sources and no consistently oriented 2-cells are created by this process. A new sink could only be created at a vertex Q which is the initial point of a dashed edge labelled c (see Fig. 4.8). ("Dashed edge" means an edge, belonging to a sink or source, that gets reversed by the above rule) Suppose such a sink occurs at Q after reversing the dashed edges in C. If only a single dashed edge labelled c is incident with Q then the remaining edges at Q must not be dashed. If there was a dashed edge labelled d incident with Q, then this edge would be pointing towards Q, and so, after reversing dashed edges, Q would not be a sink. Therefore, the cycle at Q would be of the form $b^-c^+a^-s$ where s is a path from a^- to b^- in $W^-(P)$. But, a^- and b^- lie in different components of $W^-(P)$ and hence such a path s does not exist. If several dashed edges labelled c are incident with Q and all segments $b^-c^+a^-$ of the cycle z at Q are oriented the same way, then any two such adjacent segments must be connected by a segment from a^- to b^- in $W^-(P)$ and, again, such a segment does not exist. Finally, if not all segments $b^-c^+a^-$ are oriented the same way in the cycle z then z must contain a segment of the form $b^-c^+a^-wa^-c^+b^-$ where w is a reduced nonconstant path from a^- to a^- in $W^-(P)$. w must be nonconstant, else z would not be reduced. The path w would constitute a reduced cycle in $W^-(P)$ containing a^- and we know from Fig. 4.7 that such a cycle does not exist. This proves that no new sink will be created at Q.

In the same way one proves that the edge reversals do not create new sources at points V (see Fig. 4.8) from the facts that 1. c^+ and b^+ are in different components of $W^+(P)$ and 2. there is no reduced cycle in $W^+(P)$ containing c^+. Finally, no consistently oriented region is created by the edge reversals. This can be seen directly in Fig. 4.8 which shows the four types of 2-cells adjacent to a dashed edge (that gets reversed). In order for any of these four 2-cells to become consistently oriented, additional edges would have to be reversed. The only edges (drawn solid) that may also get reversed are the edge labelled d in the left part and the edge labelled c in the right 2-cell of the right part of Fig. 4.8. In both cases, no consistently oriented 2-cell is created. In summary, the edge reversals in C would produce a graph on the 2-sphere with no consistent items, contradicting Stalling's Lemma 4.12, thus proving that $K(P)$ is DR. □

4.5 The asphericity of injective LOTs

A labelled oriented graph is called *injective* if each vertex occurs at most once as an edge label.

HUCK and ROSEBROCK have shown in [HuRo01] the following result:

Theorem 4.14 *A compressed and injective LOT which does not contain a boundary reducible sub-LOT is DR.*

HARLANDER and ROSEBROCK use this result, invent a new relative test on asphericity and use this test to show (see [HaRo17]):

Theorem 4.15 *Injective LOTs are aspherical.*

The proof of Theorem 4.14 goes roughly as follows: If the positive and the negative graph of a LOT G are trees, then G satisfies the weight test and is therefore DR. This is because one can assign weights 0 to corners of the positive or negative graph and 1 to all other edges of the Whiteheadgraph. Then the weight of each 2-cell sums up to two. Since the positive and the negative graph are trees, cycles in the Whiteheadgraph have to pass from the positive graph to the negative graph and back and so have weight at least two.

A labelled oriented graph $\mathcal{V}$ is a *reorientation* of a labelled oriented graph G if $\mathcal{V}$ is obtained from G by changing the orientation of each edge of a subset of the set of edges of G.

It is not hard to see that for a compressed injective LOT G which satisfies the weight test any reorientation of G also satisfies the weight test. This was already shown in Lemma 5.2 of [HuRo95] (see also Lemma 4.24).

HUCK and ROSEBROCK then show:

Lemma 4.16 *Let G be a compressed injective labelled oriented tree that does not contain a boundary reducible sub-LOT. Then there is a reorientation $\mathcal{V}$ of G such that $W^+(V)$ and $W^-(V)$ are trees, where V is the LOT-presentation associated with $\mathcal{V}$.*

The condition that there is no boundary reducible sub-LOT is necessary because if $\mathcal{T}$ is a labelled oriented tree that is not boundary reduced, then either $W^-(T)$ or $W^+(T)$ is not a tree. This can be seen as follows: Since $\mathcal{T}$ is not boundary reduced there exists a boundary vertex a that does not occur as an edge label. Thus, depending on the orientation of the edge of $\mathcal{T}$ containing a, either a^+ is an isolated vertex in $W^+(T)$, or a^- is an isolated vertex in $W^-(T)$. Assume without loss of generality that a^+ is isolated in $W^+(T)$. If n is the number of vertices in $\mathcal{T}$, then $W^+(T)$ contains n vertices and $n-1$ edges. Since a^+ is isolated, $W^+(T)$ contains a subgraph containing $n-1$ vertices and $n-1$

edges. Such a graph contains a cycle.

Lemma 4.16 implies Theorem 4.14 because if $\mathcal{G}$ is a compressed injective labelled oriented tree that does not contain a boundary reducible sub-LOT, then we choose the reorientation $\mathcal{V}$ which has trees as left and right graph. $\mathcal{V}$ satisfies the weight test, so $\mathcal{G}$ satisfies the weight test and is therefore DR.

The rest of this Section is devoted to a new test on relative asphericity and the proof of Theorem 4.15. It is clear that the proof of this Theorem will be harder since there are many injective LOTs which do not satisfy the weight test. We omit some details, they can be found in [HaRo17].

At first we need some notation. If $P_1 = \langle \mathbf{x}_1 \mid \mathbf{r}_1 \rangle$ and $P_2 = \langle \mathbf{x}_2 \mid \mathbf{r}_2 \rangle$ are presentations, then $P_1 \cup P_2 = \langle \mathbf{x}_1 \cup \mathbf{x}_2 \mid \mathbf{r}_1 \cup \mathbf{r}_2 \rangle$. Let $P = \langle \mathbf{x} \mid \mathbf{r} \rangle$ be a presentation and let $\{T_1, \ldots, T_n\}$ be a set of disjoint *full* sub-presentations of P. Full means that if r is a relator in P that only involves generators from T_i, then r is already a relator in T_i. Disjoint means that the generating sets of T_i and T_j are disjoint subsets of $\mathbf{x}$ in case $i \neq j$. Let $T = T_1 \cup \ldots \cup T_n$. The complex $K(T) = K(T_1) \vee \ldots \vee K(T_n)$ (the n-fold wedge product defined by identifying the vertices of the $K(T_i)$) is a sub-complex of $K(P)$.

Let $T_i = \langle \mathbf{t}_i \mid \mathbf{s}_i \rangle$ and let $\mathbf{u}_i$ be the set of cyclically reduced words with letters in $\mathbf{t}_i^{\pm 1}$ of exponent sum zero. Let $\mathbf{T}_i = \langle \mathbf{t}_i \mid \mathbf{s}_i \cup \mathbf{u}_i \rangle$. Let $\mathbf{T} = \mathbf{T}_1 \cup \ldots \cup \mathbf{T}_n$ and note that $P \cup \mathbf{T} = \langle \mathbf{x} \mid \mathbf{r} \cup \mathbf{u}_1 \cup \ldots \cup \mathbf{u}_n \rangle$ since $\mathbf{t}_i \subseteq \mathbf{x}$ and $\mathbf{s}_i \subseteq \mathbf{r}$ for every i. The presentation $P \cup \mathbf{T}$ is infinite and in the group $G(P \cup \mathbf{T})$ the generators of each T_i are identified, since we have the relator $t^{-1}t'$ in $P \cup \mathbf{T}$ for every pair t, t' of generators in T_i. Note that $\mathbf{T}_i$ is a sub-presentation of $P \cup \mathbf{T}$ and the subgraph $W(\mathbf{T}_i)$ of the Whiteheadgraph $W(P \cup \mathbf{T})$, which is spanned by the vertices $t^{\pm}$ with $t \in \mathbf{t}_i$, contains the complete graph on these vertices. In fact, every pair of vertices in $W(\mathbf{T}_i)$ is connected by infinitely many edges.

Definition 4.17 *Let Γ be a graph and Γ_0 be a subgraph. Γ is called a* forest *relative to Γ_0 if every homology reduced cycle is contained in Γ_0. Γ is called a* tree *relative to Γ_0 if Γ is connected and every homology reduced cycle is contained in Γ_0.*

The proof of the following Lemma is not very difficult (for a proof see [HaRo17]):

Lemma 4.18 *Let Γ be a graph, Γ_0 a subgraph with connected components $\Gamma_1, \ldots, \Gamma_n$. Let Γ' be the graph obtained by collapsing each component Γ_i to a vertex $g_i \in \Gamma_i$. Then Γ is a forest relative to Γ_0 if and only if Γ' is a forest.*

Definition 4.19 *Let P be a presentation, and let $\{T_1, \ldots, T_n\}$ be a set of disjoint* *full sub-presentations.* Let $T = T_1 \cup \ldots \cup T_n$. *Then* P *is said to* satisfy the Stallings test relative to T *if the following conditions hold:*

- *Relator conditions:*

  (a) Relators of $P - T$ are cyclically reduced.
  (b) Relators of $P - T$ are not positive or negative words.
  (c) Relators of T have exponent sum zero.
  (d) Any word w in the generators of some T_i that represents the trivial element of the group defined by $P \cup T$ has exponent sum zero.

- *Forest condition:*

  (e) $W^+(P \cup T)$ is a forest relative to $W^+(T)$ and $W^-(P \cup T)$ is a forest relative to $W^-(T)$.

Here is the test on relative asphericity:

Theorem 4.20 *If P is a presentation that satisfies the Stallings test relative to T, then $K(P)$ is VA relative to $K(T)$.*

Proof: Assume P satisfies the Stallings test relative to T and $K(P)$ is not VA relative to $K(T)$. Hence there exists a spherical diagram $C^* \to K(P)$ that is vertex reduced relative to $K(T)$ but does not map entirely into $K(T)$. Since $K(P)$ is a subcomplex of $K(P \cup T)$ this diagram can be viewed as a diagram $C^* \to K(P \cup T)$ which is vertex reduced relative to $K(T)$ but does not map entirely into $K(T)$. Let Ω be the collection of all spherical diagrams that have that feature. Consider the subset $\Omega_0 \subseteq \Omega$ of those diagrams for which the 2-sphere C contains the smallest number of 2-cells. From Ω_0 choose a spherical diagram $f \colon C \to K(P \cup T)$ for which C has the smallest number of edges.

It is shown in [HaRo17] that this minimality condition on C implies the following statement:

(S) If v is a vertex in C then the corner cycle $\alpha(v)$ has length at least two and does not contain two distinct corners α_p and α_q that both come from relators of one subpresentation $\mathbf{T}_i$ of $P \cup \mathbf{T}$, $i \in \{1, \ldots, n\}$.

The relator conditions (b) and (c) in Definition 4.19 (relative Stallings test) imply that C does not contain cells with consistently oriented boundary, hence C contains a sink or a source by Stallings Lemma 4.12. Let us assume without loss of generality that C contains a source, say at the vertex $v \in C$. The cycle $\alpha(v) = \alpha_1 \ldots \alpha_l$ satisfies $l \geq 2$, is contained in $W^+(P \cup \mathbf{T})$, and is homology reduced relative to $W^+(\mathbf{T})$ because $f \colon C \to K(P \cup \mathbf{T})$ is vertex reduced relative to $K(\mathbf{T})$. Since $W^+(P \cup \mathbf{T})$ is a forest relative to $W^+(\mathbf{T})$ we know that $\alpha(v)$ is

Labelled Oriented Trees and the Whitehead-Conjecture   89

entirely contained in a connected component of $W^+(\mathbf{T})$, and hence in some $W^+(\mathbf{T}_i)$, because $W^+(\mathbf{T})$ is a disjoint union of the $W^+(\mathbf{T}_i)$, $i = 1, \ldots, n$. Thus, if $\alpha(v) = \alpha_1 \ldots \alpha_l$, then all corners α_j, $j = 1, \ldots, l$ are in $W^+(\mathbf{T}_i)$. This contradicts the statement (S). We have reached a contradiction. Thus $K(P)$ is VA relative to $K(T)$. □

Let $\mathcal{P}$ be a reduced labelled oriented tree and $\{\mathcal{T}_1, \ldots, \mathcal{T}_n\}$ a set of proper maximal sub-LOTs. We assume the sub-LOTs to be pairwise disjoint, that is $\mathcal{T}_i \cap \mathcal{T}_j = \emptyset$ in case $i \neq j$. Let $\mathcal{T} = \mathcal{T}_1 \cup \ldots \cup \mathcal{T}_n$. Let P, T, T_i be the LOT-presentations associated with $\mathcal{P}$, $\mathcal{T}$, $\mathcal{T}_i$, respectively.

We denote by $\mathcal{P} - \mathcal{T}$ the forest with edge set the edges in $\mathcal{P}$ not in $\mathcal{T}$ and with vertex set the vertices which bound the edges in $\mathcal{P} - \mathcal{T}$. From each subtree $\mathcal{T}_i$ choose a vertex t_i and collapse each $\mathcal{T}_i$ in $\mathcal{P}$ to t_i to obtain a quotient tree $\mathcal{P}'$ of $\mathcal{P}$. If an edge in $\mathcal{P}'$ is labelled with a vertex $t'_i \neq t_i$ from $\mathcal{T}_i$, then relabel that edge with t_i. This turns $\mathcal{P}'$ into a labelled oriented tree. We say that $\mathcal{P}$ is *injective relative to* $\mathcal{T}$ if $\mathcal{P}'$ is injective. Note that $\mathcal{P}$ is injective relative to $\mathcal{T}$ if and only if: 1) every vertex of $\mathcal{P} - \mathcal{T}$ occurs at most once as an edge label in $\mathcal{P} - \mathcal{T}$, and 2) every $\mathcal{T}_i$ contains at most one vertex that is an edge label in $\mathcal{P} - \mathcal{T}$. It is clear that if $\mathcal{P}$ is injective itself, then $\mathcal{P}$ is also injective relative to $\mathcal{T}$. We note the following simple and useful observation.

Lemma 4.21 *If we collapse each connected component $W^+(\mathbf{T}_i)$ of $W^+(\mathbf{T})$ in $W^+(P \cup \mathbf{T})$ to the vertex t_i^+, then we obtain $W^+(P')$. The same is true if we replace W^+ by W^-.*

Example 4.22 *Figure 4.9 shows labeled oriented trees $\mathcal{P}$, $\mathcal{P}'$, $\mathcal{V}$, $\mathcal{V}'$, and the sub-LOT $\mathcal{T}$. $\mathcal{T}$ is the sub-LOT consisting of the edges labelled e, f and d in $\mathcal{P}$ and in $\mathcal{V}$. Let P, P', V, V', and T be the associated LOT-presentations, respectively. Note that $\mathcal{P}$ is reduced and injective, and the sub-LOT $\mathcal{T}$ is not boundary reduced (the boundary vertex b does not occur as an edge label in $\mathcal{T}$). Note further that $\mathcal{P}'$ is obtained from $\mathcal{P}$ by collapsing $\mathcal{T}$ to the vertex b, and that $\mathcal{V}'$ is obtained from $\mathcal{V}$ by collapsing $\mathcal{T}$ to the vertex b. The graph $W^+(P \cup \mathbf{T})$ is not a tree relative to $W^+(\mathbf{T})$, because if we collapse the connected subgraph $W^+(\mathbf{T})$ of $W^+(P \cup \mathbf{T})$ to the vertex b^+ we obtain the graph $W^+(P')$ (see Lemma 4.21), which is not a tree. It contains a 2-cycle. However we know that we can reorient $\mathcal{P}'$ to $\mathcal{V}'$, so that both $W^-(V')$ and $W^+(V')$ are trees. We now reorient $\mathcal{P}$ outside of $\mathcal{T}$ to $\mathcal{V}$. Both $W^-(V \cup \mathbf{T})$ and $W^+(V \cup \mathbf{T})$ are trees relative to $W^-(\mathbf{T})$ and $W^+(\mathbf{T})$, respectively. If we collapse the connected*

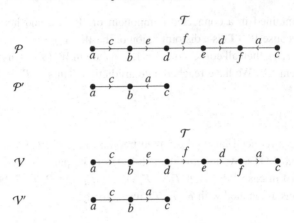

Figure 4.9 Reorienting in the presence of a sub-LOT.

subgraph $W^-(\mathbf{T})$ of $W^-(V \cup \mathbf{T})$ to the vertex b^- we obtain the graph $W^-(V')$, which is a tree; and if we collapse the connected subgraph $W^+(\mathbf{T})$ of $W^+(V \cup \mathbf{T})$ to the vertex b^+ we obtain the graph $W^+(V')$, which is also a tree.

Here is the relative version of Lemma 4.16.

Corollary 4.23 *If $\mathcal{P}$ is reduced and injective relative to $\mathcal{T}$, then there exists a reorientation $\mathcal{V}$ of $\mathcal{P}$, where only certain edges of $\mathcal{P} - \mathcal{T}$ are reoriented, so that $W^+(V \cup \mathbf{T})$ and $W^-(V \cup \mathbf{T})$ are trees relative to $W^+(\mathbf{T})$ and $W^-(\mathbf{T})$, respectively. In fact, V satisfies the Stallings test relative to T.*

Proof: Consider the quotient LOT $\mathcal{P}'$ obtained from $\mathcal{P}$ by collapsing each $\mathcal{T}_i$ to a vertex t_i. The LOT $\mathcal{P}'$ is compressed since $\mathcal{P}$ is reduced.

Theorem 4.16 implies that there is a reorientation $\mathcal{V}'$ of $\mathcal{P}'$, such that $W^+(V')$ and $W^-(V')$ of the LOT-presentation V' of $\mathcal{V}'$ are trees. Let $\mathcal{V}$ be a reorientation of $\mathcal{P}$, where no edge of $\mathcal{T}$ is reoriented (so $\mathcal{T}$ is contained in $\mathcal{V}$), so that collapsing each $\mathcal{T}_i$ in $\mathcal{V}$ to the vertex t_i results in $\mathcal{V}'$. Let P, P', V, V', T, and T_i be LOT-presentations associated with $\mathcal{P}, \mathcal{P}', \mathcal{V}, \mathcal{V}', \mathcal{T}$, and $\mathcal{T}_i$, respectively. Collapsing the components $W^+(\mathbf{T}_i)$ of $W^+(\mathbf{T})$ in $W^+(V \cup \mathbf{T})$ to t_i^+ yields the tree $W^+(V')$ (see Lemma 4.21). So it follows from Lemma 4.18 that $W^+(V \cup \mathbf{T})$ is a forest relative to $W^+(\mathbf{T})$. Since $W^+(V')$ is connected, so is $W^+(V \cup \mathbf{T})$. Hence $W^+(V \cup \mathbf{T})$ is a tree relative to $W^+(\mathbf{T})$. In the same way we can argue that $W^-(V \cup \mathbf{T})$ is a tree relative to $W^-(\mathbf{T})$.

This shows that the forest condition (e) of Definition 4.19 (relative Stallings test) holds. The relator conditions (a)-(d) also hold because V is a LOT-presentation that comes from the reduced LOT $\mathcal{V}$. So, all relators of V are cyclically reduced and have exponent sum zero. This implies that all relators of $V \cup \mathbf{T}$ have exponent sum zero and hence each word in the generators of V that represents the trivial element in the group defined by $V \cup \mathbf{T}$ has exponent sum zero. □

Let $\mathbf{x}$ be a set and w be a word in $\mathbf{x}^{\pm 1}$. Let S be a subset of $\mathbf{x}$. Define w_S to be the word obtained from w by replacing x^ϵ in w by $x^{-\epsilon}$, $\epsilon = \pm 1$, if and only if $x \in S$. If $\mathbf{w}$ is a set of words in $\mathbf{x}^{\pm 1}$, then let $\mathbf{w}_S$ be the set of words w_S, $w \in \mathbf{w}$. If $P = \langle \mathbf{x} \mid \mathbf{r} \rangle$ is a presentation, denote by $P_S = \langle \mathbf{x} \mid \mathbf{r}_S \rangle$.

The map $x \to x^\epsilon$, where x is a generator in P and $\epsilon = 1$ if x is not in S and $\epsilon = -1$ if x is in S, results in a homeomorphism $\phi \colon K(P) \to K(P_S)$ on the corresponding standard 2-complexes. Furthermore, if $f \colon C \to K(P)$ is a vertex reduced spherical diagram, then so is $\phi \circ f \colon C \to K(P_S)$. In particular both P and P_S present the same group. So, if P is a LOT-presentation, then P_S is also a presentation of a LOT-group.

Let $\mathcal{P}$ be a reduced labelled oriented tree and let P be the associated LOT-presentation. Let S be a subset of the generators of P. Let $\mathcal{V}$ be the reorientation of $\mathcal{P}$ where exactly those edges are reoriented which have their label in S. Let V be the LOT-presentation associated with $\mathcal{V}$.

The following lemma is essentially Lemma 5.2 of [HuRo95].

Lemma 4.24 *The Whiteheadgraphs $W(P_S)$ and $W(V)$ are equal.*

Proof: This follows since the relators $r_e = xzy^{-1}z^{-1}$ and $(r_e)_S = x^{-1}zy^{-1}z^{-1}$ contribute the same edges to the Whiteheadgraphs, which can be seen in Figure 4.10. □

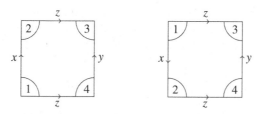

Figure 4.10 The relators $r_e = xzy^{-1}z^{-1}$ and $(r_e)_S = x^{-1}zy^{-1}z^{-1}$.

We assume as before that $\mathcal{P}$ is a reduced LOT and $\{\mathcal{T}_1, \ldots, \mathcal{T}_n\}$ is a set of proper maximal pairwise disjoint sub-LOTs. Let $\mathcal{T} = \mathcal{T}_1 \cup \ldots \cup \mathcal{T}_n$. Let P, T, T_i be the LOT-presentations associated with $\mathcal{P}$, $\mathcal{T}$, $\mathcal{T}_i$, respectively.

Theorem 4.25 *If $\mathcal{P}$ is injective relative to $\mathcal{T}$, then there exists a subset S of the generators in P so that P_S satisfies the Stallings test relative to T_S.*

Proof: By Corollary 4.23 there exists a reorientation $\mathcal{V}$ of $\mathcal{P}$ so that both $W^{\pm}(V \cup \mathbf{T})$ are trees relative to $W^{\pm}(\mathbf{T})$. Only edges in $\mathcal{P}-\mathcal{T}$ change orientation. Let S_0 be the subset of generators of P that occur as labels on edges that change orientation when passing from $\mathcal{P}$ to $\mathcal{V}$. We enlarge S_0 to a set $S = S_0 \cup \bigcup_{i=1}^n S_i$, where S_i is the empty set if no generator of T_i is contained in S_0, and S_i is the set of generators of T_i if S_0 contains a generator from T_i. Let $\mathcal{V}^*$ be the reorientation of $\mathcal{P}$ where exactly the edges with labels in S are reoriented. Note that $\mathcal{V}^*$ contains a reorientation $\mathcal{T}^*$ of $\mathcal{T}$, and that $\mathcal{V} - \mathcal{T} = \mathcal{V}^* - \mathcal{T}^*$. Since $\mathbf{T} = \mathbf{T}^*$ this implies that $V \cup \mathbf{T} = V^* \cup \mathbf{T}^*$. Thus $W(V \cup \mathbf{T}) = W(V^* \cup \mathbf{T}^*)$, and so both $W^{\pm}(V^* \cup \mathbf{T}^*)$ are trees relative to $W^{\pm}(\mathbf{T}^*)$.

We first show that the forest condition (e) of Definition 4.19 holds for the pair $T_S \subseteq P_S$. Note that

$$P \cup \mathbf{T} = \langle \mathbf{x} \mid \mathbf{r} \cup \bigcup_{i=1}^n \mathbf{u}_i \rangle,$$

where $P = \langle \mathbf{x} \mid \mathbf{r} \rangle$ and $\mathbf{u}_i$ is the set of cyclically reduced words of exponent sum zero in the generators of T_i. Thus

$$P_S \cup \mathbf{T_S} = \langle \mathbf{x} \mid \mathbf{r}_S \cup \bigcup_{i=1}^n \mathbf{v}_i \rangle,$$

where $\mathbf{v}_i$ is the set of cyclically reduced words of exponent sum zero in the generators of $(T_i)_S$. Since the generating sets of T_i and $(T_i)_S$ are the same, we have that $\mathbf{v}_i = \mathbf{u}_i$ for all i. So

$$P_S \cup \mathbf{T_S} = \langle \mathbf{x} \mid \mathbf{r}_S \cup \bigcup_{i=1}^n \mathbf{u}_i \rangle.$$

Now,

$$V^* \cup \mathbf{T}^* = \langle \mathbf{x} \mid \mathbf{z} \cup \bigcup_{i=1}^n \mathbf{w}_i \rangle,$$

where $V^* = \langle \mathbf{x} \mid \mathbf{z} \rangle$ and $\mathbf{w}_i$ is the set of cyclically reduced words of exponent sum zero in the generators of T_i^*. Since the generating sets of T_i and T_i^* are the same, we have that $\mathbf{w}_i = \mathbf{u}_i$ for all i. So

$$V^* \cup \mathbf{T}^* = \langle \mathbf{x} \mid \mathbf{z} \cup \bigcup_{i=1}^{n} \mathbf{u}_i \rangle.$$

Since $W(P_S) = W(V^*)$ by Lemma 4.24 and the same set of relators $\mathbf{u} = \bigcup_{i=1}^{n} \mathbf{u}_i$ is added when enlarging P_S to $P_S \cup \mathbf{T_S}$ as when enlarging V^* to $V^* \cup \mathbf{T}^*$, it is clear that $W(P_S \cup \mathbf{T_S}) = W(V^* \cup \mathbf{T}^*)$. Since both $W^{\pm}(V^* \cup \mathbf{T}^*)$ are trees relative to $W^{\pm}(\mathbf{T}^*)$, and $W(\mathbf{T}^*) = W(\mathbf{T_S})$, it follows that both $W^{\pm}(P_S \cup \mathbf{T_S})$ are trees relative to $W^{\pm}(\mathbf{T_S})$. Thus, the forest condition holds.

We have to check the relator conditions (a)-(d). (a), (b) and (c) is clear, we show (d):

Let w be a word in the generators of some $(T_i)_S$ that represents the trivial element in the group defined by $P_S \cup \mathbf{T_S} = \langle \mathbf{x} \mid \mathbf{r}_S \cup \bigcup_{i=1}^{n} \mathbf{u}_i \rangle$, where $P = \langle \mathbf{x} \mid \mathbf{r} \rangle$ and $\mathbf{u}_i$ is the set of words of exponent sum zero in the generators of T_i (see above). Then w_S is a word in the generators of T_i that represents the trivial element in the group defined by $\langle \mathbf{x} \mid \mathbf{r} \cup \bigcup_{i=1}^{n} \mathbf{u}_{iS} \rangle$. Since by construction S contains either no generator of T_i or all of them, it follows that $\mathbf{u}_{iS} = \mathbf{u}_i$ for all i. So w_S is a word in the generators of T_i that represents the trivial element in the group defined by $P \cup \mathbf{T} = \langle \mathbf{x} \mid \mathbf{r} \cup \bigcup_{i=1}^{n} \mathbf{u}_i \rangle$. Since all relators of $P \cup \mathbf{T}$ have exponent sum zero, w_S has exponent sum zero. Thus $w_S \in \mathbf{u}_i$ and hence $w \in \mathbf{u}_{iS} = \mathbf{u}_i$. In particular w has exponent sum zero. □

We are now ready to give the proof of Theorem 4.15:

Proof: We proceed by induction on the number of vertices. If $\mathcal{P}$ consists of a single vertex, then the result is true.

If $\mathcal{P}$ is not reduced, then we transform it into a reduced injective labelled oriented tree $\mathcal{P}_{red}$ that contains fewer vertices than $\mathcal{P}$. Thus, by induction hypothesis, $\mathcal{P}_{red}$ is aspherical, and hence so is $\mathcal{P}$, because $K(\mathcal{P})$ is homotopically equivalent to $K(\mathcal{P}_{red})$.

From now on we assume that $\mathcal{P}$ is reduced. If $\mathcal{P}$ contains no proper sub-LOTs, then Theorem 4.14 gives the desired result.

So, assume there are proper sub-LOTs. Let $\{\mathcal{T}_1, \ldots, \mathcal{T}_n\}$ be the set of maximal proper sub-LOTs of $\mathcal{P}$. Note that every $\mathcal{T}_i$ is compressed and injective and contains fewer vertices than $\mathcal{P}$. Hence, by induction, each $\mathcal{T}_i$ is aspherical.

Case 1. Suppose that for some i, j we have $\mathcal{T}_i \cap \mathcal{T}_j \neq \emptyset$.

We assume without loss of generality that $\mathcal{T}_1 \cap \mathcal{T}_2 \neq \emptyset$. Then, by maximality of the sub-LOTs it follows that $\mathcal{P} = \mathcal{T}_1 \cup \mathcal{T}_2$. Note that there might be more than two $\mathcal{T}_i$. The intersection $\mathcal{T}_{12} = \mathcal{T}_1 \cap \mathcal{T}_2$ is a sub-LOT. Indeed, if b is an edge label in $\mathcal{T}_{12}$, then b has to be a vertex of $\mathcal{T}_1$, because $\mathcal{T}_1$ is a sub-LOT, and b has to be a vertex of $\mathcal{T}_2$, because $\mathcal{T}_2$ is a sub-LOT. Thus b is a vertex of $\mathcal{T}_{12}$. The authors show in [HaRo17] by induction that $\pi_1(K(\mathcal{T}_{12})) \to \pi_1(K(\mathcal{T}_i))$, $i = 1, 2$, is injective. So $\pi_1(K(\mathcal{P})) = \pi_1(K(\mathcal{T}_1)) *_{\pi_1(K(\mathcal{T}_{12}))} \pi_1(K(\mathcal{T}_2))$ is an amalgamated product. Furthermore, since both $K(\mathcal{T}_i)$ and the intersection $K(\mathcal{T}_{12})$ are aspherical by induction hypothesis, and the inclusion induced maps $\pi_1(K(\mathcal{T}_{12})) \to \pi_1(K(\mathcal{T}_i))$, $i = 1, 2$, are injective, a theorem of Whitehead [Whi39] implies that $K(\mathcal{P})$ is aspherical as well.

Case 2. The $\mathcal{T}_i$, $i = 1, \ldots, n$, are pairwise disjoint.

Let P, T_i, and T be the LOT-presentations of $\mathcal{P}$, $\mathcal{T}_i$, and $\mathcal{T}$, respectively. By Theorem 4.25 there exists a subset S of the set of generators of P so that P_S satisfies the Stallings test relative to T_S. It follows from Theorem 4.20 that $K(P_S)$ is VA relative to $K(T_S)$. Using the homeomorphism $\phi : K(P) \to K(P_S)$ defined above Lemma 4.24 we conclude that $K(P)$ is VA relative to $K(T)$. By induction hypothesis $\pi_2(K(T_i)) = 0$ and hence $\pi_2(K(T)) = 0$. Now $\pi_2(K(P)) = 0$ follows from Theorem 4.4.

The diagram in Figure 4.11 provides an overview of the situation in case 2. □

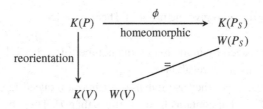

Figure 4.11 Overview of the proof

4.6 Two examples

We first show in an example how to prove asphericity for a non-injective LOT which satisfies the Stallings test relative to a sub-LOT.

Example 4.26 *Let $\mathcal{P}$ be the LOT of Figure 4.12. Let $\mathcal{T}$ be the sub-LOT spanned by the vertices x_1, x_2, x_3, x_4, x_5. $K(\mathcal{T})$ is aspherical by Theorem 4.15 although its positive and negative graph both contain cycles. $\mathcal{P}$ is reduced and not injective. $\mathcal{P}$ satisfies the Stallings test relative to $\mathcal{T}$ since the relative positive and negative graphs are trees. So $K(P)$ is VA relative to $K(T)$ by Theorem 4.20 and hence $K(P)$ is aspherical by Theorem 4.4.*

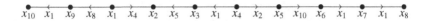

Figure 4.12 A LOT satisfying the relative Stallings test.

The second example is an example of a non-aspherical LOG with certain properties. If $f: C \to K(P)$ is a spherical diagram over a 2-complex $K(P)$ modeled on a LOG-presentation P, then there is the induced map $f_*: L \to \mathcal{G}$ which maps the dual link into the corresponding LOG $\mathcal{G}$ (see Section 4.3). If $\mathcal{G}$ is a LOT then $f_*(L)$ is contractible since in a tree all closed paths are contractible.

We describe in this example a LOG $\mathcal{G}$ and a surjective reduced spherical diagram $f: C \to K(\mathcal{G})$ (found with the help of a computer program written by the author) with the following properties:

1 $K(\mathcal{G})$ has the Euler-Characteristic of a LOT-complex,
2 all paths $f_*(L)$ are contractible in $\mathcal{G}$,
3 the π_2-element defined by $[f]$ is non-trivial.

Example 4.27 *The LOG $\mathcal{G}$ of Figure 4.13 admits the reduced spherical diagram C of Figure 4.14. For simplicity in Figure 4.14 arcs between undercrossing and undercrossing are labelled by i instead of x_i. The two paths in Figure 4.14 lift to contractible paths in $\mathcal{G}$. C corresponds to a non-trivial π_2-element because the only two 2-cells of Figure 4.14 corresponding to the relator $x_1 x_4 = x_4 x_5$ are in different levels of the covering space corresponding to the map $\bar{f}: G \to \mathbb{Z}$ which sends each x_i to 1.*

The author has written software which runs in parallel on several hundred computers and checks LOTs for reduced spherical diagrams. The program has been running since 2008 and has checked millions of LOTs and billions of diagrams. It has found several hundreds of reduced spherical diagrams, non

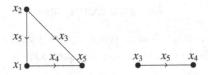

Figure 4.13 A LOG with non-trivial second homotopy group.

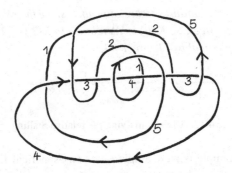

Figure 4.14 A corresponding spherical diagram

of which are non-trivial π_2-elements. Typically LOTs which admit reduced spherical diagrams seem to present knot-groups, when containing a boundary-reducible sub-LOT, or the infinite cyclic group. A LOT $\mathcal{T}$ is *minimal* if it has the minimal number of vertices among all LOTs with LOT group isomorphic to $G(\mathcal{T})$.

Conjecture: *Minimal LOTs are DR.*

If minimal LOTs are DR, then all LOTs are aspherical. Because if K and L are 2-complexes with the same fundamental group and the same Euler-characteristic and K is aspherical, then so is L.

4.7 Virtual Knots and LOTs

Virtual knots where defined by KAUFFMANN in [Kau99]. A *virtual link diagram* is a planar 4-regular graph (a graph where each vertex has valency 4) with under- and over crossing information at some but not necessarily all nodes. A virtual link diagram can be thought of as a regular projection of a link in $S \times I$

to S where S is an orientable surface. A virtual knot diagram is a virtual link diagram with only one link component.

A *labelled oriented circle* (LOC) is a LOG, where the underlying graph is a circle and a *labelled oriented interval* is a LOG where the underlying graph is an interval. HARLANDER and ROSEBROCK observed in [HaRo10] that the class of virtual knot groups agrees with the class of LOC groups. In the same paper they show:

Theorem 4.28 *Let $\mathcal{G}$ be a LOC and assume that $K(\mathcal{G})$ is aspherical. Then $G(\mathcal{G})$ is a virtual knot group but not a knot group.*

Proof: $H_2(K(\mathcal{G})) = H_2(G(\mathcal{G}))$ since $K(\mathcal{G})$ is aspherical. One can construct a non-trivial element of $H_2(K(\mathcal{G}))$ for a LOC $\mathcal{G}$ by grouping the 2-cells of $K(\mathcal{G})$ in a row achieving a torus. On the other hand it is well known (see BURDE/ZIESCHANG [I], [BuZi85]), that $H_2(J) = 0$ if J is a knot group. □

A *long virtual knot diagram k* is obtained by cutting a virtual knot diagram at a point on an edge, thus producing a graph that has exactly two nodes of valency one. A Wirtinger presentation $P(k)$ can be read off in the usual way. It is easy to see that $P(k) = P(\mathcal{P})$, where $\mathcal{P}$ is a labelled oriented interval. More details on the connection between labelled oriented intervals and long virtual knots can be found in [HaRo03]. We say a long virtual knot diagram is *aspherical*, if the standard 2-complex associated with the Wirtinger presentation is aspherical. A virtual knot diagram is *alternating* if one encounters over- and under-crossings in an alternating fashion when traveling along the diagram. A long alternating virtual knot diagram is obtained when cutting an alternating virtual knot diagram. The following is a direct consequence of Theorem 4.15.

Corollary 4.29 *A long alternating virtual knot diagram k is aspherical.*

Proof: The labelled oriented interval that records the Wirtinger presentation of k is injective. The result follows from Theorem 4.15. □

4.8 L^2-homology and Whitehead's asphericity conjecture

By Jens Harlander

The results discussed in this section are due to HILLMAN [Hil97] (see also Chapter 12, written by BERRICK and HILLMAN, in "Guido's Book of Conjectures", edited by CHATTERJI [BeHi08]), OSIN-THOM[OsTh13] and OSIN [Osi15]). Standard references for L^2-homology are ECKMANN [Eck00] and LÜCK [Lüc01].

Let G be a finitely generated (and hence countable) group. Recall that the group algebra $\mathbb{R}G$ is the vector space of finite formal sums over G with coefficients in $\mathbb{R}$. An element of $\mathbb{R}G$ is written as a sum $\sum_{g \in G} f(g)g$, $f(g) \in \mathbb{R}$, and $f(g) = 0$ for all but finitely many $g \in G$. Denote by $l^2(G)$ the Hilbert space of square-summable formal sums over G with coefficients in $\mathbb{R}$. An element of $l^2(G)$ is represented as a sum $\sum_{g \in G} f(g)g$, $f(g) \in \mathbb{R}$, $\sum_{g \in G} f(g)^2 < \infty$. The inner product on $l^2(G)$ is defined by

$$\left\langle \sum_{g \in G} f(g)g, \sum_{g \in G} h(g)g \right\rangle = \sum_{g \in G} f(g)h(g).$$

The set G is a vector space basis for $\mathbb{R}G$ and G is an orthonormal Hilbert space basis (a maximal orthonormal subset) for $l^2(G)$. The space $l^2(G)$ is the Hilbert space completion of $\mathbb{R}G$. The group G acts on $l^2(G)$ by left-multiplication:

$$h \sum_{g \in G} f(g)g = \sum_{g \in G} f(g)hg = \sum_{g \in G} f(gh^{-1})g.$$

However, $l^2(G)$ is not a ring.

If V is a Hilbert space, then we call $W \subseteq V$ a *Hilbert subspace* if W is a closed linear subspace with induced Hilbert space structure. A *Hilbert G-module* is a left G-module M, which is a Hilbert space on which G acts by isometries such that M is isometrically G-isomorphic to a G-invariant Hilbert subspace of $l^2(G)^n$ for some n.

The *von Neumann algebra* $N(G)$ is the algebra of bounded (left) G-invariant operators $l^2(G) \to l^2(G)$. Note that

$$\mathbb{R}G \subseteq N(G) \subseteq l^2(G).$$

The *trace* of an element $\phi \in N(G)$ is defined as

$$\mathrm{tr}_G(\phi) = \langle \phi(1), 1 \rangle.$$

Using the trace one can define the *von Neumann dimension* $\dim_G(M)$ of a Hilbert G-module M. The von Neumann dimension satisfies the following basic properties:

1. $\dim_G(M) \geq 0$, and $\dim_G(M) = 0$ if and only if $M = 0$;
2. if $\dim_G(M) = \dim_G(N)$, then $M \cong N$;
3. $\dim_G(M \oplus N) = \dim_G(M) + \dim_G(N)$;
4. $\dim_G(M) \leq \dim_G(N)$ in case $M \subseteq N$;
5. $\dim_G(l^2(G)) = 1$

Let X be a CW-complex with fundamental group G, $\tilde{X}$ its universal covering, and $(C_*(\tilde{X}, \mathbb{R}), \partial)$ the associated cellular chain complex. Note that each $C_i(\tilde{X}, \mathbb{R})$ is a free G-module of the form $\mathbb{R}G^{m_i}$, where m_i is the number of i-cells in X. We apply the functor $l^2(G) \otimes_{\mathbb{R}G} -$ to the chain complex $(C_*(\tilde{X}, \mathbb{R}), \partial)$ and obtain the chain complex $(C_*^{(2)}(\tilde{X}), \partial^{(2)})$. Each $C_i^{(2)}(\tilde{X})$ is a Hilbert G-module of the form $l^2(G)^{m_i}$, where m_i is the number of i-cells in X. Let $K_i = \ker \partial_i^{(2)}$, and $I_i = \operatorname{im} \partial_{i+1}^{(2)}$. Define the ith L^2-*homology* of X to be

$$H_i^{(2)}(X) = K_i / \bar{I}_i,$$

where $\bar{I}_i$ denotes the closure. The ith-L^2 *betti-number* is defined as

$$\beta_i^{(2)}(X) = \dim_G(H_i^{(2)}(X)).$$

One can also do this for groups instead of CW-complexes. Let G be a group. Define

$$\beta_i^{(2)}(G) = \beta_i^{(2)}(X),$$

where X is any $K(G, 1)$-complex. One can show that $\beta_0^{(2)}(G) = \frac{1}{|G|}$ in case G is finite, and $\beta_0^{(2)}(G) = 0$ in case G is infinite. Furthermore, if X is a 2-complex and G its fundamental group, then $\beta_j^{(2)}(X) = \beta_j^{(2)}(G)$, $j = 0, 1$, because a $K(G, 1)$-complex can be obtained by attaching cells to X.

It turns out that L^2-Betti numbers, just like the ordinary Betti numbers, can be used to compute the Euler-characteristic for finite complexes. If X is a finite CW-complex of dimension n, then

$$\chi(X) = \sum_{i=1}^{n} (-1)^i \beta_i^{(2)}(X).$$

Lemma 4.30 *Let X be a finite 2-complex and assume $\chi(X) = 0$. Let $G = \pi_1(X)$. The following statements are equivalent:*

1. $\beta_1^{(2)}(G) = 0$;
2. $H_2^{(2)}(X) = 0$.

Proof: Since $\chi(X) = 0$ the group G is infinite and hence $\beta_0^{(2)}(G) = 0$. Thus
$$0 = \chi(X) = \beta_0(G) - \beta_1^{(2)}(G) + \beta_2^{(2)}(X) = 0 - \beta_1^{(2)}(G) + \beta_2^{(2)}(X).$$
So, $\beta_1^{(2)}(G) = \beta_2^{(2)}(X)$. It follows that $\beta_2^{(2)}(X) = 0$ if and only if $\beta_1^{(2)}(G) = 0$. □

Corollary 4.31 *Let X be a 2-complex with fundamental group G and Euler-characteristic $\chi(X) = 0$. If $\beta_1^{(2)}(G) = 0$, then X is aspherical.*

Proof: By Lemma 4.30 we have $H_2^{(2)}(X) = 0$. This implies that the boundary map
$$C_2^{(2)}(\tilde{X}) \xrightarrow{\partial_2^{(2)}} C_1^{(2)}(\tilde{X})$$
is injective. If we restrict $\partial_2^{(2)}$ to $C_2(\tilde{X})$ it becomes the boundary map
$$C_2(\tilde{X}) \xrightarrow{\partial_2} C_1(\tilde{X}).$$
Thus ∂_2 is injective and hence $\pi_2(X) = 0$. □

The other direction in the above corollary is false. We can produce an aspherical 2-complex X with $\chi(X) = 0$ so that $\beta_1^{(2)}(G) \neq 0$, where G is the fundamental group of X. Let $Y = (S^1 \vee S^1) \times (S^1 \vee S^1)$, so the fundamental group of Y is of the form $F(a,b) \times F(c,d)$. Let $X = Y \vee S^1$ and let G be the fundamental group of X. Then X is aspherical, $\chi(X) = 0$, but $\beta_1^{(2)}(G) \neq 0$. This follows from the product (Künneth) formula for L^2-Betti numbers:
$$\beta_n^{(2)}(A \times B) = \sum_{i+j=n} \beta_i^{(2)}(A)\beta_j^{(2)}(B).$$
We have $\beta_0^{(2)}(S^1 \vee S^1) = 0, \beta_1^{(2)}(S^1 \vee S^1) = 1$, and $\beta_i^{(2)}(S^1 \vee S^1) = 0$, for $i \geq 2$. Thus, $\beta_2^{(2)}(Y) = 1$. Since $\tilde{Y} \subseteq \tilde{X}$ we have $\beta_2^{(2)}(X) \neq 0$ and hence we also have $\beta_1^{(2)}(G) \neq 0$ by Theorem 4.30.

Theorem 4.32 *Let X be a finite 2-complex and assume that we can attach a single 2-cell to X to obtain a contractible 2-complex Y. Let G be the fundamental group of X. If $\beta_1^{(2)}(G) = 0$ then X is aspherical.*

Proof: Since Y is contractible we have $\chi(Y) = 1$ and hence $\chi(X) = 0$. The result follows from Corollary 4.31. □

Corollary 4.33 *Let $\mathcal{T}$ be a labelled oriented tree such that $\beta_1^{(2)}(G(\mathcal{T})) = 0$. Then $\mathcal{T}$ is aspherical.*

Various criteria that guarantee the vanishing of the L^2-Betti numbers are known. For a list we refer to Lück [Lüc01], Theorem 7.2. The following is a direct consequence of item (7) on that list: If G is a finitely generated group and N is a finitely generated normal subgroup of G such that G/N is infinite, then $b_1^{(2)}(G) = 0$.

Define the *normal rank* nrk(G) of a finitely generated group G to be the minimal number of elements that normally generate G. In [OsTh13] Osin and Thom provide evidence that the first L^2-Betti number gives a lower bound for the normal rank of a group. They state the following conjecture.

Conjecture: *(Osin-Thom rank conjecture) Let G be a finitely generated torsion free group. Then*

$$\beta_1^{(2)}(G) \leq \text{nrk}(G) - 1.$$

Note that the group G considered in Theorem 4.32 has normal rank 1. So if the conjecture is true, it follows that a subcomplex X of a contractible 2-complex Y that is obtained from X by attaching a single 2-cell, is aspherical in case $\pi_1(X) = G$ is torsion free. In particular the following result holds.

Theorem 4.34 *Let $\mathcal{T}$ be a labelled oriented tree and assume that $G(\mathcal{T})$ is torsion free. If the Osin-Thom rank conjecture is true, then $\mathcal{T}$ is aspherical.*

An isometric action of a group G on a metric space S is called *acylindrical* if for every $\epsilon > 0$ there exist $R, N > 0$ such that for every two points $x, y \in S$ satisfying $d(x, y) \geq R$, there are at most N elements $g \in G$ such that $d(x, gx) \leq \epsilon$ and $d(y, gy) \leq \epsilon$. A group G is called *acylindrically hyperbolic* if it admits a non-elementary acylindrical action on a hyperbolic metric space.

The class of acylindrically hyperbolic groups includes non-elementary hyperbolic and relatively hyperbolic groups. The following result is due to Osin [Osi15].

Theorem 4.35 *Let G be a finitely presented group and assume that*

1 $\beta_1^{(2)}(G) > 0$;
2 some finite index subgroup of G maps onto $\mathbb{Z}$.

Then G is acylindrically hyperbolic.

Corollary 4.36 *Let $\mathcal{T}$ be a labelled oriented tree. Assume that $G(\mathcal{T})$ is not acylindrically hyperbolic. Then $\mathcal{T}$ is aspherical.*

Proof: The group $G = G(\mathcal{T})$ satisfies the second condition of the above theorem. It cannot satisfy the first condition because it is not acylindrically hyperbolic. Thus $\beta_1^{(2)}(G) = 0$ and the result follows from Corollary 4.33. □

This reduces the asphericity question for labelled oriented trees to the ones that have acylindrically hyperbolic LOT groups. This prompts the following question:

Question: Suppose $\mathcal{T}$ is a labelled oriented tree for which $G(\mathcal{T})$ is non-elementary hyperbolic. Is $\mathcal{T}$ aspherical?

5
2-Complexes and 3-Manifolds

Janina Glock, Cynthia Hog-Angeloni and Sergei Matveev

5.1 Introduction

A 2-dimensional spine K^2 of a 3-manifold M^3 carries all (simple) homotopy information of M^3, see Chapter I in [I]. In case of a closed 3-manifold this holds after the interior of a 3-ball has been removed and collapses have been performed. Starting with a 2-dimensional complex the question arises whether there exists an embedding into a 3-manifold and if so, whether the regular neighbourhood $N(K^2)$ of K^2 is unique up to homeomorphism.

In Section 5.2 we take up the questions about existence and uniqueness of 3-manifold thickening posed as questions (48) and (49) in Chapter I in [I]. In [I], page 263 and 266, algorithms are given to decide whether a standard 2-complex is a spine of a 3-manifold. In section 5.2 we show three types of obstructions to uniqueness and prove that imposing corresponding restrictions on the 3-manifold into which we embed, the regular neighbourhood of K^2 is unique.

In order to find conditions for uniqueness based on the 2-complex (rather than on the manifold), the notion of k-connectivity of the Whitehead graph is discussed in section 5.3. We will investigate the question:

To what extent can the structure of the fundamental group of a 2-complex be seen in its Whitehead graph, provided that the 2-complex is a 3-manifold spine?

Section 5.4 - 5.7 deal with prime decompositions of various geometric objects.

The famous Kneser-Milnor Theorem states that any closed orientable 3-manifold can be decomposed into a connected sum of prime manifolds, which admit no further decomposition, see [Kne29], [Mil62]. Also, any knot in the 3-sphere is a connected sum of prime knots, see [Sch49] which cannot be decomposed into a nontrivial connected sum. In both cases, the prime factors are unique up to reordering. The idea to study geometric objects by decomposing them into smaller pieces is very natural. We describe here a general method for proving prime decomposition theorems of different kinds: for global knots, for spatial graphs, for orbifolds. Two examples are discussed in this chapter: the Kneser-Milnor prime factorization theorem for closed connected orientable 3-manifolds (Section 6) and knots in $F \times I$ where F is a closed orientable surface (Section 7).

The method is based on a generalization of the Diamond Lemma by M.H.A. Newman [New42], which turned out to be very useful in various fields of mathematics, especially in algebra, see the work of G. M. Bergman [Ber78], and the theory of Gröbner bases [BeWe99]. It is called the Diamond Lemma, since its proof is based on a rhombus-shaped diagram. This lemma was brought to our attention by the work of Stefanie Zentner [Zen05].

Our goal is to present several results on existence and uniqueness of prime decompositions of different geometric objects. The upshot is that all main results are obtained by the same universal scheme. The variative parts of the proofs consist in applying different old and new methods of removing intersections of surfaces, which have been well-elaborated in the last century.

5.2 Existence and Uniqueness

In this chapter we assume all complexes and manifolds to be compact and piecewise linear (p.l.); all manifolds should be orientable, all 2-complexes connected.

Following the questions which have been stated in Chapter I in [I] we give an overview of (partial) results.

Question (48): *Does a given 2-dimensional complex embed into some 3-manifold?*

In general the answer is no, but Neuwirth's algorithm determines the answer for any given standard complex; see [I], [Ne68]. Osborne and Stevens ([I], [OsSt74], [OsSt77]) derived relator conditions and an improved algorithm for 2-generator presentations and the second author ([I], [Ho-An92]) extended their results to any finite number of generators.

In case an embedding does exist, the following question arises:

2-Complexes and 3-Manifolds

Question (49): *Are regular neighbourhoods inside a given 3-manifold uniquely determined by K^2?*

In general the answer is no, as the following Example 5.1 illustrates.

Example 5.1 *Consider two embeddings f_1, f_2 of $K^2 = S^2 \vee S^1$ in $S^2 \times S^1$, where the map f_1 attaches S^1 to one side of S^2, and f_2 connects both sides of S^2 with S^1 (see Figure 5.1). Then the regular neighbourhood $N(f_1(K^2))$ has two boundary components and $N(f_2(K^2))$ only one. Thus $N(f_1(K^2))$ and $N(f_1(K^2))$ are not homeomorphic.*

Figure 5.1 Two ways to thicken $S^2 \vee S^1$

As already mentioned in [I], page 35 and 36, Casler [I], [Ca65] has shown that every compact, connected 3-manifold has a spine which is a special polyhedron and conversely, that the thickening of a special polyhedron is unique. Here,

Definition 5.2 *A 3-manifold M^3 is a thickening of a 2-complex K^2, if $K^2 \subset M^3$ and $M^3 \searrow K^2$.*

see [I] page 31.

Casler's result is stronger than asked in question (49), as it is independent of the ambient manifold.

The first two authors [GlHo05] investigated Question (49) for more general spines (of particular interest are standard spines of presentations) and collected three types of examples for non-uniqueness:

- The boundary of the regular neighborhood of K^2 is not connected.
  E.g. see example 5.1 above.
- M is a connected sum.
  E.g. The lens spaces $L_{5,1}$ and $L_{5,2}$ share the same spine K^2, but they are non-homeomorphic. Let $M = L_{5,1} \# L_{5,2}$ then K^2 has two embeddings in M with non-homeomorphic regular neighborhoods.

- A regular neighborhood of K^2 contains essential annuli.

  E.g. The square and the granny knot are connected sums of two copies of the trefoil knot. Their knot complements in S^3 are not homeomorphic, as there is no isomorphism of their knot groups which carries the peripheral subgroup of one to the peripheral subgroup of the other, but they share the same spine, see [Fox52].

  (Here, the peripheral subgroup of a knot group is the subgroup of the fundamental group generated by the boundary torus.)

Excluding these three types of obstructions, it is shown in [GlHo05], Theorem 3.2:

Theorem 5.3 *Let M be a compact, connected, orientable, prime 3-manifold. If $f_1 \colon K^2 \to M$ and $f_2 \colon K^2 \to M$ are embeddings into the interior of M, $N_1 = N(f_1(K^2))$ and $N_2 = N(f_2(K))$ are regular neighbourhoods in M and both N_1 and N_2 do not contain essential annuli and have connected boundary, then N_1 and N_2 are homeomorphic.*

Note that, as the double of any 3-manifold has Euler-characteristic equal to 0, $\chi(\partial N) = 2\chi(N) = 2\chi(K^2)$ and thus under the assumption of Theorem 5.3, the homeomorphism type of the boundary of $N(f(K^2))$ is determined by K^2.

The proof of Theorem 5.3 uses results from 3-manifold theory such as JSJ-decomposition and theorems of Waldhausen and Johannson. A related result has been proved by M. Burdon [Bur01]:

Theorem 5.4 *Let M be a closed, connected 3-manifold, and let two homotopic embeddings of a 2-complex K^2 into M be given. Assume that the boundaries of the regular neighbourhoods of the embedded 2-complex are both homeomorphic to the 2-sphere. Then the regular neighbourhoods are homeomorphic.*

5.3 k-Connectivity of Whitehead Graphs

Let $F = F(X)$ be a free group on a finite set $X = \{x_1, ..., x_k\}$, let $R = \{r_1, ..., r_\ell\}$ be a set consisting of cyclically reduced words in F and let K^2 be the standard complex with just one vertex v of the presentation $\mathcal{P} = \langle X \mid R \rangle$, see [I, page 9].

Assume $\mathcal{P}$ is a 3-manifold presentation, i. e. the corresponding 2-complex embeds into some 3-manifold M. Then the intersection of the boundary of a small regular neighbourhood of v (which is a 2-sphere) with the standard complex itself is a planar graph, called the *Whitehead graph* $\mathcal{W}(\mathcal{P})$ of $\mathcal{P}$. It contains a pair of vertices for each generator of $\mathcal{P}$, and its edges correspond to *syllables of length 2* in some cyclically written defining relator of $\mathcal{P}$, i. e. sequences of two adjacent symbols $x_i^{\pm 1}$, $x_j^{\pm 1}$.

Of course, the Whitehead graph is defined for an arbitrary presentation but throughout this chapter we only deal with 3-manifold presentations and thus with planar Whitehead graphs.

The Whitehead graph is also known as *link graph* whereas in the *reduced Whitehead graph* or *star graph*, multiple edges of the Whitehead graph are represented by a single edge (see [I], page 170).

A *simple graph* is a graph with neither loops nor multiple edges. The reduced Whitehead graph of $\mathcal{P}$ is simple because the relators of $\mathcal{P}$ are assumed to be cyclically reduced.

W. METZLER has asked whether from the Whitehead graph of a 3-manifold presentation $\mathcal{P}$, we can obtain information about

- uniqueness of thickenings ([I], Question (49))
- the structure of the underlying group.

The Whitehead graph $\mathcal{W}(\mathcal{P})$ together with its embedding into S^2 carries all the information about the 3-manifold. The generators of $\mathcal{P}$ give rise to 1-handles which are attached to the regular neighborhood of v and the relations of $\mathcal{P}$ give rise to 2-handles which are attached to the resulting handlebody. Since the 1- and 2-handles can be attached in exactly one way, the following holds:

Lemma 5.5 *If the embedding of the Whitehead graph $\mathcal{W}(\mathcal{P})$ of a given presentation $\mathcal{P}$ of a 2-complex K^2 into S^2 is unique up to isotopy, then K^2 has a unique thickening.* □

Hence, if we want to get uniqueness for thickenings of a given 3-manifold presentation we look for conditions under which the embedding of $\mathcal{W}(\mathcal{P})$ in S^2 is unique up to isotopy.

Connectivity is one of the basic concepts of graph theory and plays a crucial role here; it asks for the minimum number of vertices that need to be removed to disconnect the graph.

Definition 5.6 *A graph G is called n-separated if it is the union of two subgraphs G_1 and G_2 with the following properties:*

- *no common edges*
- *the number of common vertices is smaller or equal to n*
- *each G_i has a vertex not belonging to the other*

A graph G which is not n-separated is called (n + 1)-connected. The connectivity number $\kappa(G)$ is defined as the largest k such that G is k-connected.

Note that $\kappa(G)$ is the only integer for which G is both k-separated and k-connected.

Example 5.7 *The graph G on the left in Figure 5.2 is 3- and 2-separated, illustrated in Figure 5.2 in the middle resp. on the right, but is not 1-separated. G is n-separated for $n \geq 2$, n-connected for $n \leq 2$ and has connectivity number $\kappa(G) = 2$.*

Figure 5.2 3- and 2-separability of a graph with connectivity number $\kappa(G) = 2$

A graph with connectivity number 0 is just a disconnected graph. A graph with connectivity number 1 has a *cut vertex*, also known as *articulation point* (see [I], page 347), which is the common vertex of the Γ_i. For a graph Γ with connectivity number 2, we shall call two distinct vertices which lie in the intersection of the graphs Γ_i a *pair of cut vertices*.

The following result by Whitney mainly asserts uniqueness of thickenings in case of a 3-connected Whitehead graph, see the subsequent discussion.

Theorem 5.8 *Every 3-connected, simple, planar graph has a unique embedding (up to isotopy) in S^2.*

The proof can be found in ([Whi32b],[Whi33a]).

Planarity of the Whitehead graph follows from the assumption that $\mathcal{P}$ be a 3-manifold presentation. For a 3-connected Whitehead graph, it follows from Theorem 5.8 that an embedding of the reduced Whitehead graph into S^2 is

2-Complexes and 3-Manifolds 109

unique up to isotopy.

Regarding the assumption "simple", for the moment we work with the reduced Whitehead graph; at a later stage we need to analyze the consequences of the existence of multiple occurrences of the same syllable of length 2 (resp. multiple edges) to uniqueness of the thickening. This is a topic of further investigation. For a small example showing an ambiguity arising from multiple edges in the Whitehead graph see the lens space example before Theorem 5.3. The Whitehead graph of both $L_{5,1}$ and $L_{5,2}$ consists of 2 vertices and 5 edges stemming from the common spine with just one cell in dimensions 0, 1 and 2 respectively.

Note that there are 0-, 1- and 2-connected graphs, which have non-isotopic embeddings into S^2, for a visualization of the latter two see Figure 5.3.

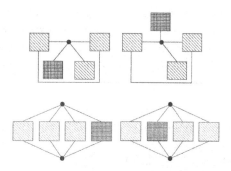

Figure 5.3 Non-uniqueness of planar embeddings of 1-connected (to the left) and 2-connected (to the right) graphs. Each square contains some complicated portion of the graph

Hence, we are now going to investigate Whitehead graphs with connectivity number 0-, 1- or 2.

John Stallings [Sta99] has related non-connectedness of the Whitehead graph of a presentation $\mathcal{P}$ to a decomposition of the underlying group as a free product.

Definition 5.9 *A presentation $\mathcal{P}$ is called* separated *if for some presentations $\mathcal{P}_1, \mathcal{P}_2$ it has the form $\mathcal{P}_1 \sqcup \mathcal{P}_2$ (disjoint union).*

Note that if $\mathcal{P}$ is separated and the groups presented by $\mathcal{P}_1, \mathcal{P}_2$ are nontrivial then $\mathcal{P}$ is a presentation of a free product.

Every pair of vertices in $\mathcal{W}(\mathcal{P})$, which belongs to the same generator, has the same *valency* (number of adjacent edges), corresponding to the occurrences of this generator in the relators. If an element of X does not occur in any relator of $\mathcal{P}$, then the corresponding two vertices within $\mathcal{W}(\mathcal{P})$ have zero valency; such vertices are said to be *isolated*. Note that if $\mathcal{W}(\mathcal{P})$ contains an isolated vertex, then R is separated; in fact, the underlying group splits off a $\mathbb{Z}$-factor.

More generally, Stallings [Sta99] showed:

Theorem 5.10 *Assume that $\mathcal{W}(\mathcal{P})$ is not connected; $\mathcal{W}(\mathcal{P}) = \Gamma_1 \sqcup \Gamma_2$ with nonempty, disjoint graphs Γ_1, Γ_2.*

1 If the sets of vertices of Γ_1 resp. Γ_2 are closed under inverse, meaning that for any element of X the two corresponding vertices of $\mathcal{W}(P)$ are either both contained in Γ_1 or both contained in Γ_2, then $\mathcal{P}$ is separated.
2 If not, then a Whitehead transformation can be performed in a way that $\mathcal{W}(\mathcal{P})$ contains an isolated vertex after the basis change and thus becomes separated.

Here, a *Whitehead transformation* is a change of basis of the free group $F(X)$ that corresponds to a different choice of meridians of the handlebody, which arises as a regular neighborhood of the 1-skeleton of K^2, see the book of Lyndon and Schupp ([I], [LySc77]).

Proof: The first statement is obvious.

For the second statement, look at the regular neighbourhood of the vertex $v \in K^2$, see figure 5.4, with the Whitehead graph $\mathcal{W}(\mathcal{P}) = \Gamma_1 \sqcup \Gamma_2$ on its boundary. In the figure, the vertices of $\mathcal{W}(\mathcal{P})$ have been thickened to small discs which come in pairs $x_i, \bar{x}_i$; when these are identified for all i, we obtain the regular neighbourhood of the 1-skeleton of the 2-complex corresponding to $\mathcal{P}$.

Let c be a simple closed curve separating Γ_1 and Γ_2 and assume that x lies in Γ_1 and $\bar{x}$ in Γ_2. Perform a Whitehead transformation substituting x by c (by cutting the ball in Figure 5.4 along a disc bounded by c and identifying x and $\bar{x}$). No relator curve crosses c, so $\mathcal{P}$ splits off a factor $\langle c \mid - \rangle$. □

Thus the connectivity 0 case explains the second type of obstruction before Theorem 5.3: Connected sums have a spine with 0-connected Whitehead

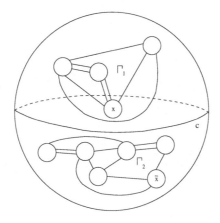

Figure 5.4 Whitehead graph of connectivity $k = 0$.

graph.

The case $k = 1$ turns out to be a degenerate case:

Theorem 5.11 [folklore] *Assume that $W(\mathcal{P})$ has a cut vertex. Then there is a Whitehead transformation decreasing the sum of the lengths of the relators.*

Proof: As in the proof of Theorem 5.10 we look at the regular neighbourhood of the vertex $v \in K^2$, see figure 5.5, with the Whitehead graph $W(\mathcal{P}) = \Gamma_1 \cup \Gamma_2$ on its boundary, where now $\Gamma_1 \cap \Gamma_2$ consists of just one vertex. Again, in figure 5.5 the vertices of $W(\mathcal{P})$ have been thickened to small discs. Let b_1 be an arc with endpoints on the disc corresponding to the generator x separating Γ_1 and Γ_2, let b_2 be an arc with same endpoints as b_1 on the boundary of that disc x such that if we push $b_1 \cup b_2$ slightly to the side, we get a simple closed curve c separating x and $\bar{x}$, see fig. 5.5.

Perform a Whitehead transformation substituting x by c (by cutting the ball along a disc bounded by c and identifying x and $\bar{x}$). Note that the total length of the relators has decreased by the number of edges from x to vertices of Γ_1.

□

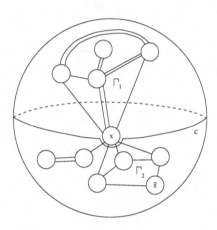

Figure 5.5
Whitehead graph of connectivity $k = 1$

This case can thus be reduced to the others.

It remains to investigate **2-connected** Whitehead graphs $\mathcal{W}(\mathcal{P})$, which breaks into two subcases according to whether the pair of cut vertices belongs to different resp. to the same generator.

Let us discuss a sample case of a 2-connected Whitehead graph, the pair of cut vertices belonging to the same generator x, see Fig. 5.6.

Assume that Γ_1 and Γ_2 are closed under inverse, and that there is a band Q inside the handle corresponding to the generator x which is disjoint from the relator curves and separates the endpoints of the edges at x and $\bar{x}$ that lie in Γ_1 from the endpoints of those in Γ_2.

Then the band Q together with a disc D bounded by a curve c separating Γ_1 and Γ_2, see Figure 5.6, forms an annulus A properly embedded into the handlebody avoiding the relator curves.

Definition 5.12 *A presentation* $\mathcal{P} = \langle X \mid R \rangle$ *is called* $\mathbb{Z}$*-separated if it has the form* $\mathcal{P}_1 \cup \mathcal{P}_2$ *where* $\mathcal{P}_1 = \langle x_1, \ldots, x_\ell \mid S \rangle$ *and* $\mathcal{P}_2 = \langle x_\ell, \ldots, x_n \mid T \rangle$, $n \geq \ell+1 \geq 3$, $S \cup T = R$.

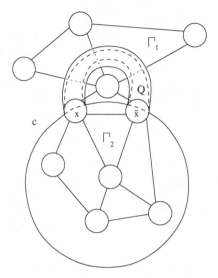

Figure 5.6 The annulus $A = Q \cup D$ is the union of a band Q inside the handle corresponding to the generator x and a disc D inside the 3-ball which is a regular neighbourhood of the vertex v

Denote by G_i the group presented by $\mathcal{P}_i$, $i = 1, 2$. Note that if $\mathcal{P}$ is $\mathbb{Z}$-separated with the cyclic subgroups generated by x_ℓ in G_1 resp. G_2 proper and infinite, then $\mathcal{P}$ is a presentation of a free product amalgamated over an infinite cyclic group.

We are currently exploring whether for connectivity number equal to 2 there always is a Whitehead transformation such that after the corresponding basis change $\mathcal{P}$ is either $\mathbb{Z}$-separated or has shorter length.

This sheds some light on the third type of obstruction mentioned before Theorem 5.3: A 3-manifold M^3 containing an essential, separating annulus A possesses a spine K^2 with Whitehead graph having connectivity number 2. Here, (the base point of K^2 and) the pair of cut vertices can be thought of as being contained in A while all the other vertices and (open) edges of the Whitehead graph are contained in the complement of A.

5.4 M.H.A. Newman's Diamond Lemma and a new version of it

Definition 5.13 *A binary relation* $\to$ *on a set X is* antisymmetric *if there is no pair of distinct elements $a, b \in X$ such that $a \to b$ and $b \to a$.*

More formally, $\to$ is antisymmetric precisely if for all a and b in X if $a \to b$ and $b \to a$, then $a = b$. This relation is *well-founded* if every non-empty subset $S \subset X$ has a *terminal* element, that is, some element $m \in X$ such that m is not related to any element of S.

A binary relation $\to$ on X is called *confluent* if for any $a, b, c \in X$ such that $a \to b$ and $a \to c$ there is $d \in X$ such that $b \to \to d$ and $c \to \to d$, where $\to \to$ denotes the reflexive transitive closure of $\to$.

Lemma 5.14 *[Newman's Diamond Lemma] [New42]. Let $\to$ be a confluent well-founded binary relation on a set X. Then any element of X has a unique terminal element.*

As we have indicated in the introduction, instead of using the terminology of binary relations, we prefer an equivalent language borrowed from the theory of graphs, see Fig. 5.7.

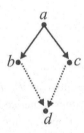

Figure 5.7 Graphical illustration of the confluent condition and the diamond lemma.

Let Γ be an arbitrary oriented graph. We do not impose any condition on Γ. It may be infinite and have infinite valences of vertices. We denote its vertex and edge sets by $V(\Gamma)$ and $E(\Gamma)$, respectively. An oriented path in Γ is a sequence $\{(\overrightarrow{A_1A_2}, \overrightarrow{A_2A_3}, \ldots)\}$ of oriented edges in which the beginning of each edge coincides with the end of the previous one.

2-Complexes and 3-Manifolds

Definition 5.15 *[Ho-AnMa08] We say that a vertex $B \in V(\Gamma)$ is a root of a vertex $A \in V(\Gamma)$ if*

1 there is an oriented path in Γ from A to B;
2 B is terminal, that is, it does not have outgoing edges.

A vertex of Γ may have no roots, one root, or several roots. See Fig. 5.8.

Question: *Under what conditions on Γ does each of its vertices have exactly one root?*

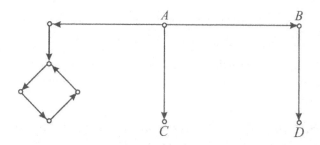

Figure 5.8 A has two roots, B only one root, C, D are their own roots. All other vertices have no roots

We formulate two properties of Γ. The first is called the *Finite Paths property* and is denoted by (FP):

(FP): Any oriented path in Γ is finite.

In other words Γ does not contain oriented cycles or infinite oriented paths. Note that (FP) implies that each vertex A of Γ has at least one root.

Denote by $E^2(\Gamma)$ the set of all unordered pairs of edges of the form $(\overrightarrow{AB_1}, \overrightarrow{AB_2})$ with $B_1 \neq B_2$, that is, pairs of edges with a common beginning and different endpoints. We now formulate the second property (MF) called *Mediator Function*.

(MF): There exists a map $\mu : E^{(2)}(\Gamma) \to \mathbb{N} \cup \{0\}$, called the mediator function, which satisfies the following conditions:

(MF1) if $\mu(\overrightarrow{AB_1}, \overrightarrow{AB_2}) = 0$, then there exist oriented paths from B_1 and from B_2 ending at the same vertex $C \in V(\Gamma)$;

(MF2) if $\mu(\overrightarrow{AB_1}, \overrightarrow{AB_2}) > 0$ then there exists an edge $\overrightarrow{AC}$ such that $\mu(\overrightarrow{AB_i}, \overrightarrow{AC}) < \mu(\overrightarrow{AB_1}, \overrightarrow{AB_2})$ for $i = 1, 2$.

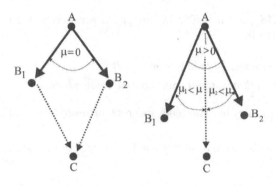

Figure 5.9 Graphical illustration of the new Diamond Lemma.

The following lemma holds for arbitrary graphs. It is a reminiscent of the famous Newman Diamond Lemma 5.14, but is much more suitable for geometric applications.

Lemma 5.16 *(Diamond Lemma) If an oriented graph Γ has the properties (FP) and (MF), then each of its vertices has a unique root.*

Proof: The existence of a root follows from property (FP). We proceed to prove uniqueness. We call a vertex *regular* if it has exactly one root, and *singular* otherwise. Similarly, an edge is *regular* if and only if its endpoint is regular. Arguing by contradiction, assume that Γ has a singular vertex. Then there exists a singular vertex A such that all its outgoing edges are regular. Otherwise each singular vertex has a singular outgoing edge, and therefore every path with a singular end can be continued further by adding new singular edges, in contradiction to (FP).

Choose two edges $\overrightarrow{AB_1}$ and $\overrightarrow{AB_2}$ such that B_1 and B_2 have different roots and the number $\mu(\overrightarrow{AB_1}, \overrightarrow{AB_2})$, which we denote μ_0, is minimal possible among all numbers $\mu(\overrightarrow{AX_1}, \overrightarrow{AX_2})$ with X_1 and X_2 having different roots. Note that B_1, B_2 are regular and their roots are different.

Consider the two cases $\mu_0 = 0$ and $\mu_0 > 0$. In the first case any root of a vertex C in (MF1) is a common root for B_1 and B_2. Since both these vertices are regular, they do not have other roots, which contradicts the choice of B_1 and B_2.

In the second case we obtain a contradiction between the property (MF2) and the choice of the edges $\overrightarrow{AB_1}$ and $\overrightarrow{AB_2}$. Indeed, let C be the end of the edge $\overrightarrow{AC}$ existing by (MF2). Since all vertices B_1, B_2 and C are regular, each of them has a unique root. The roots of B_1 and B_2 are different, and therefore at least one of the roots (say, the root of B_1) is different from the root of C. Since $\mu(\overrightarrow{AB_1}, \overrightarrow{AC}) < \mu_0$, this contradicts the minimality of μ_0. □

5.5 Universal Scheme

The proofs of all results in the subsequent sections of this chapter follow the same pattern. Let O be the class of objects under consideration. Usually these objects are compact connected 3-manifolds, or pairs of the form (M, G) where M is a compact connected manifold and G is a compact one-dimensional polyhedron, for example, a knot inside M. We first construct a graph Γ whose vertices are finite sets of objects in O. By passing to the disjoint union of all the objects at a given vertex, we may consider this vertex as a single not necessarily connected manifold, or as a single pair of type (M, G) with possibly disconnected M and G.

The edges of Γ are defined by means of non-trivial reductions. Each reduction is a surgery of an object $x \in$ O. The first step is to cut this object along an orientable surface S embedded in it; then we apply a gluing. The reduction therefore assigns one or two objects in O to x, depending on whether the surface S is separating. The set of reductions under consideration depends on the nature of the problem.

Reductions of objects define reductions of vertices of Γ, and therefore define its edges. Two vertices A, B of Γ are joined by the edge $\overrightarrow{AB}$ if they consist of the same objects with one exception: one object in A undergoes a reduction, which replaces it by a new object or pair of objects in B. The same edge may correspond to different reductions. An edge is determined by the initial and terminal sets of objects only, and does not depend on the particular way in which a reduction is carried out. Hence Γ does not contain multiple edges.

The proof of property (FP) is usually straightforward although there are exceptions. By contrast, the mediator function in (MF) is always constructed in the same way. Its value $\mu(\overrightarrow{AB_1}, \overrightarrow{AB_2})$ at a pair of edges is defined as the minimal number #$(S_1 \cap S_2)$ of connected components of the intersection of surfaces $S_1, S_2 \subset A$ defining the edges $\overrightarrow{AB_1}, \overrightarrow{AB_2}$ respectively, where the minimum is taken over all pairs of such surfaces. Of course, here we assume that surfaces are in general position. If $\mu(\overrightarrow{AB_1}, \overrightarrow{AB_2}) = 0$ that is, if the edges are defined by disjoint surfaces, then each of these surfaces survives reduction along the other one. This implies that the vertex obtained as a result of both reductions does not depend on which of the reductions is carried out first. Therefore, it can be taken as the vertex C in the formulation of (MF1).

Assume now that $\mu(\overrightarrow{AB_1}, \overrightarrow{AB_2}) > 0$. This implies B_1, B_2 are different, since otherwise μ would not be defined.

In order to establish property (MF2) it is enough to show that for any two surfaces $S_1, S_2 \subset A$ defining the edges $\overrightarrow{AB_1}$, $\overrightarrow{AB_2}$ there exists a surface $S \subset A$ defining an edge $\overrightarrow{AC}$ such that $\mu(\overrightarrow{AB_i}, \overrightarrow{AC}) < \mu(\overrightarrow{AB_1}, \overrightarrow{AB_2})$ for $i = 1, 2$.

We refer to this S as a mediator surface. The remarkable observation is that practically in all cases the standard technique of removing intersections is sufficient for finding a mediator surface. This technique has long been used in low-dimensional topology and is quite well developed. Of course, the particular methods for constructing mediator surfaces depend on the nature of the problem. Nevertheless, the universal scheme described above allows us either to achieve the desired result, that is, to prove uniqueness of a root for each vertex, or to identify precisely the main obstacle, which usually leads to negative results, i.e. to constructing counterexamples to uniqueness. In the next two sections we describe two examples of using the universal scheme.

5.6 Kneser-Milnor Theorem

As a first application of the Diamond Lemma, we give a new proof of the Kneser-Milnor Theorem on prime decompositions of 3-manifolds. Let S be a separating sphere inside a closed orientable 3-manifold M. Recall that a *spherical reduction* of M along S consists in cutting M along S and gluing up by balls the two copies of S in the boundary of the resulting manifolds. We get two new closed 3-manifolds M_1, M_2. In this case it is said that M is a *connected sum* $M_1 \# M_2$.

Definition 5.17 *A 3-manifold $M \neq S^3$ is called* prime *if it admits no nontrivial decomposition into a connected sum.*

Theorem 5.18 (Kneser-Milnor prime decomposition theorem) *[Kne29], [Mil62]. Any closed connected orientable 3-manifold M different from S^3 is homeomorphic to a connected sum of prime 3-manifolds. The summands are defined uniquely up to reordering.*

Proof: For simplicity we only consider the case when M does not contain non-separating 2-spheres (in this case there are no summands $S^1 \times S^2$). We follow the universal scheme and start with constructing the graph Γ. The set $V(\Gamma)$ of its vertices consists of all compact orientable 3-manifolds, possibly

2-Complexes and 3-Manifolds

disconnected. Two vertices A, B of Γ are connected by an oriented edge $\overrightarrow{AB}$ if B is obtained from A by a nontrivial reduction along a sphere S contained in a connected component of A. Thus B contains one manifold more than A.

According to the universal scheme, for proving the theorem, it suffices to show that Γ has properties (FP) and (MF). Property (FP) is a direct consequence of the famous Lemma of Kneser [Kne29]. In fact, the number of nontrivial spherical reductions of any 3-manifold M is bounded by 14Δ, where Δ is the number of tetrahedra in a triangulation of M.

In order to prove that Γ possesses property (MF) we follow the universal scheme and construct the mediator function $\mu : E^{(2)}(\Gamma) \to \mathbb{N} \cup \{0\}$ as follows. The value $\mu(\overrightarrow{AB_1}, \overrightarrow{AB_2})$ at a pair of edges is defined as the minimal number $\#(S_1 \cap S_2)$ of circles in the intersection of spheres $S_1, S_2 \subset A$ which define the edges $\overrightarrow{AB_1}, \overrightarrow{AB_2}$. The minimum is taken over all pairs of such spheres.

Let us prove that μ satisfies the conditions (MF1),(MF2). We first consider the case $\mu(\overrightarrow{AB_1}, \overrightarrow{AB_2}) = 0$, i.e. the spheres S_1, S_2 are disjoint. Then each of them survives reduction along the other one. Therefore we may assume that $S_1 \subset B_2$ and $S_2 \subset B_1$. Reducing B_1 along S_2 and B_2 along S_1, we get the same vertex C. This proves property (MF1).

In order to prove property (MF2) we consider the case when two spheres defining the edges $\overrightarrow{AB_1}, \overrightarrow{AB_2}$ have nonempty intersection, that is, $\mu_0 = \mu(\overrightarrow{AB_1}, \overrightarrow{AB_2}) > 0$. Then these spheres lie in the same connected manifold $Q \subset A$. According to the universal scheme it suffices to show that there exists a mediator sphere, i.e. a non-trivial sphere $S \subset A$ satisfying $\#(S_i \cap S) < \#(S_1 \cap S_2)$ for $i = 1, 2$. Any nontrivial sphere in $A \setminus Q$ is a mediator. If such a sphere does not exist we construct a mediator sphere by surgery. Among all circles in $S_1 \cap S_2$ we choose one, denoted by c, which is innermost with respect to S_1. This circle bounds a disc D in S_1 such that $D \cap S_2 = c$. We cut S_2 along c and glue up the boundaries of the cut by two parallel copies of D. After a small perturbation we get two new spheres S' and S'' which do not intersect S_2 and intersect S_1 in a smaller number of circles, see Fig. 5.10.

At least one of these two spheres (say, S') is non-trivial in Q since otherwise S_2 would be trivial. Thus S' is a mediator sphere. It follows that Γ admits a mediator function and we can apply the Diamond Lemma. □

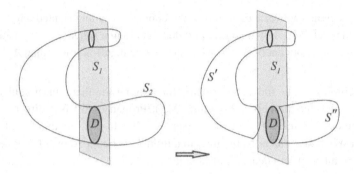

Figure 5.10 surgery along an innermost circle

The proof of the general case when there are non-separating spheres is slightly more complicated and can be found in [Mat12b].

5.7 Prime decompositions of knots in $F \times I$

Let F be a closed orientable surface and $I = [0, 1]$.

Definition 5.19 *A knot in $F \times I$ is a simple closed curve in the interior of $F \times I$. Two knots $K \subset F \times I$ and $K' \subset F' \times I$ are equivalent if there exists a homeomorphism $F \times I \to F' \times I$ taking K to K'.*

Knots of the above type are interesting since many properties of classical knots carry over to knots in thickened surfaces. This is not surprising since the theory of knots in S^3 is equivalent to the theory of knots in $S^2 \times I$. Like knots in S^3, knots in thickened surfaces can be represented by diagrams similar to diagrams of classical knots. The only difference is that they lie in F, not in a plane or a sphere, as in the classical case. Reidemeister moves play the same role: they realize isotopies of knots.

Let $F \times I$ be a thickened surface. We will say that a proper annulus $A \subset F \times I$ is *vertical* if it is isotopic to an annulus of the form $c \times I$ where c is a simple closed curve in F. Consider two knots $K_i \subset F_i \times I$, $i = 1, 2$. In order to define their annular connected sum $K_1 \# K_2 \subset (F_1 \# F_2) \times I$, we choose discs $D_i \subset F_i$ and deform each K_i isotopically so that the intersection $l_i = K_i \cap (D_i \times I)$ becomes a trivial proper arc in the ball $D_i \times I$.

Definition 5.20 *An annular connected sum of knots $K_1 \subset F_1 \times I$ and $K_2 \subset F_2 \times I$ is a knot $K = K_1 \# K_2$ in $F = (F_1 \# F_2) \times I$ obtained by gluing the manifolds*

$(F_i \setminus \mathrm{Int} D_i) \times I$ *together by a homeomorphism* $\varphi \colon \partial D_1 \times I \to \partial D_2 \times I$ *such that* $\varphi(\partial l_1) = \partial l_2$ *and* $\varphi(\partial D_1 \times \{0\}) = \partial D_2 \times \{0\}$.

The last condition guarantees that the orientation of I doesn't get inverted.

At first glance this definition has little in common with the definition of a connected sum of classical knots. However, if it is reformulated in the language of diagrams, then we obtain the same definition as in the classical case. Indeed, we cut from the given surfaces F_1 and F_2 a pair of discs D_1, D_2 intersecting the knot diagrams in a single simple arc each, and then glue the resulting surfaces together by a homeomorphism φ of their boundaries preserving the endpoints of the above arcs. See Fig. 5.11.

Figure 5.11 Annular connected sum of two knots in thickened surfaces.

In the case $F_1, F_2 = S^2$ we obtain the connected sum operation in the classical sense. A connected sum of knots in thickened surfaces depends on the choice of the discs D_1 and D_2, isotopical deformations of knots, and the homeomorphism φ. Therefore, $K_1 \# K_2$ is a multivalued operation. In general, the number of different connected sums of two given knots is infinite even in the case when the knots are trivial, that is, bound discs. See Fig. 5.12.

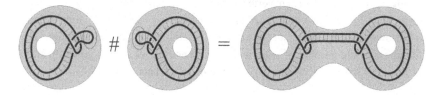

Figure 5.12 Annular connected sum of two trivial knots in a thickened surface can produce many different nontrivial knots.

The inverse operation to annular summation is called *annular reduction*. It is performed as follows. First we cut $F \times I$ and the given knot $K \subset F \times I$ along a separating vertical annulus intersecting K in two points. We get two thickened surfaces. The boundary of each of them contains one of the annuli created by the cut and each one contains an arc of the knot K whose endpoints lie in the corresponding annulus. Then we glue up those annuli by two index two handles, i.e. by thickened discs with trivial proper arcs of K inside them such that the endpoints of one arc are glued to the endpoints of the other one.

We construct the corresponding oriented graph Γ. Its vertices are sets of knots in thickened surfaces. Vertices which differ by addition or deletion of trivial knots in $S^2 \times I$ are considered equivalent. Edges correspond to nontrivial reductions along vertical annuli and spheres.

We also allow auxiliary reductions such as cutting $F \times I$ along separating vertical annuli disjoint from the knot. Parts of the given manifold which contain no knots are discarded. Spherical reductions along spheres intersecting K in two points are also allowed. (In fact, these can already be obtained via an annular reduction along a vertical annulus corresponding to a simple closed curve in F that bounds a disc.)

Note that if one of the summands in $K_1 \# K_2$ is an unknot in $S^2 \times I$, then the knot $K_1 \# K_2$ is equivalent to the second summand. A connected sum of this form is said to be *trivial*.

A knot in $F \times I$ is said to be *prime*, if it cannot be represented as a non-trivial connected sum of two other knots.

Theorem 5.21 *If a knot $K \subset F \times I$ is nontrivial, i.e. does not bound an embedded disc in $F \times I$, then it is either prime or can be decomposed into a connected sum of prime knots.*

Proof: We assign to each vertex V of Γ its *weight* $w(V) = \Sigma_i g(F_i)^2$, with knots $K_i \subset F_i \times I$. Each annular reduction produces a vertex of smaller weight. It follows that any sequence of successive annular reductions is finite. The finiteness of the number of spherical reductions follows from a generalization of Kneser's lemma (see [Kne29]) and the proof of Theorem 5.18. □

Of course, this also follows from additivity of knot genus.

Theorem 5.21 says nothing about uniqueness of the prime summands. The reason is that uniqueness does not hold. We provide the appropriate counterexamples below (see Fig. 5.17). It turns out that uniqueness does hold if we restrict ourselves to a smaller class of knots, see Theorem 5.23.

Definition 5.22 *A knot $K \subset F \times I$ is said to be homologically trivial if it defines a trivial element in the homology group $H_1(F \times I; \mathbb{Z}_2) = H_1(F; \mathbb{Z}_2)$.*

2-Complexes and 3-Manifolds   123

It is easy to see that the class of homologically trivial knots is closed with respect to connected sums.

The following theorem is an analogue (for knots in thickened surfaces) of the Schubert Theorem [Sch49] on prime decompositions of classical knots.

Theorem 5.23 *If a homologically trivial knot is decomposed into a connected sum of prime summands, then these summands are determined uniquely up to permutation.*

Proof: We shall follow the universal scheme.

Property (FP) has been established in Theorem 5.21.

Property (MF) can be proved in the same way as in the general scheme and in the proof of the Kneser-Milnor theorem. The only additional argument needed here is to show how one may decrease the number of circles and radial segments (arcs) in the intersections of two vertical annuli (this is needed for getting property (MF2)). Note that the intersection of two annuli may only consist of trivial circles and arcs, non-trivial circles, and radial arcs. Since the annuli are embedded, non-trivial circles and radial arcs cannot occur together. Trivial circles and arcs are removed by innermost circle and outermost arc surgery. Nontrivial circles can be removed as follows:

Let A_1, A_2 be two vertical annuli such that $A_1 \cap A_2$ consists of nontrivial circles. The mutual positions of these annuli is indicated by arcs a_1, a_2 in a rectangle, see Fig. 5.13. Think of the annuli A_1, A_2 to be obtained by rotating the arcs and rectangle around the dotted axis.

The mediator annulus we are looking for can be obtained by rotating the arc a_3. This arc runs along a_2 up to its last intersection with a_1, but instead of crossing a_1 it turns down to the boundary of the rectangle.

Consider the remaining situation when the intersection of two given annuli A_1, A_2 consists of disjoint radial arcs. Then the pair $(F \times I, A_1 \cup A_2)$ is vertical, that is, has the form $(F \times I, (c_1 \cup c_2) \times I)$, where c_1, c_2 are circles in F. Therefore, we can forget about the annuli and the given knot K, and consider only the circles c_1, c_2 and the projection $\bar{K}$ of K. This reduces the problem of removing intersections between annuli to the one between circles. This can be solved by standard techniques and a few ad hoc-tricks. We follow the universal scheme but instead of performing surgeries along surfaces in 3-manifolds we perform surgeries along circles in surfaces.

Assume that $c_1 \cap c_2$ consists of $k \geq 5$ points. They divide each circle into k arcs. Since $k \geq 5$, one of the circles (say, c_1) contains a pair α, β of adjacent arcs which do not intersect $\bar{K}$. We apply surgery to c_2 along α by cutting c_2

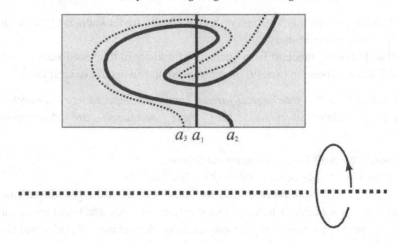

Figure 5.13 The mediator annulus is indicated by the dotted line in the rectangle.

at the points in $\partial\alpha$ and attaching two parallel copies of the arc α, see Figure 5.14. Since the curve c_2 separates F, we obtain two circles c', c'', which are either both separating, or both non-separating. In the first case one of the circles c', c'' is a mediator. What shall we do in the second case? This situation can be resolved by applying a surgery to the union $c' \cup c''$ along an arc which connects c', c'' and is parallel to the arc β (see Fig. 5.14).

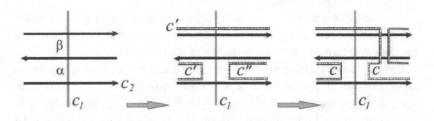

Figure 5.14

As a result, we obtain a non-trivial separating circle $c \subset F$ intersecting the projection $\bar{K}$ at two points corresponding to the points in $c_2 \cap \bar{K}$. It can be taken as a mediator. Indeed, $\#(c \cap c_1) < \#(c_1 \cap c_2)$ since two points in the intersection $c_1 \cap c_2$ have disappeared, and $\#(c \cap c_2) < \#(c_1 \cap c_2)$, because $\#(c \cap c_2) = 4$, and $\#(c_1 \cap c_2) \geq 5$ by assumption.

So far, we haven't even used that the knot be homologically trivial.

Since the circles c_1 and c_2 are both separating, they intersect in an even number of points. Therefore, this intersection can consist either of two or four points. The first possibility is easy, see [Mat12b] for details.

Let us consider the more interesting and conceptual second possibility, when $\mu(\overrightarrow{AB_1}, \overrightarrow{AB_2}) = \#(c_1 \cap c_2) = 4$. We will consider $c_1 \cup c_2$ as a graph with 4 vertices and 8 edges forming 4 biangles.

Case 1. Suppose that $\bar{K}$ crosses each edge at not more than one point. Since K is homologically trivial, the following holds: if $\bar{K}$ crosses an edge of a biangle at one point, then it must cross the other edge of this biangle also at one point. Taking into account that $\bar{K}$ crosses each circle at two points, we get two possible ways how $\bar{K}$ can intersect a regular neighborhood $N = N(c_1 \cup c_2)$, see Fig. 5.15 (a), (b), where intersections of $\bar{K}$ and N are shown as dotted segments.

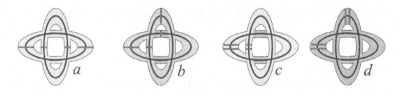

Figure 5.15 All possible ways how $\bar{K}$ can intersect N

Case 2. Suppose that there are no arcs intersecting $\bar{K}$ at one point, but there is an arc intersecting $\bar{K}$ at two points. In this situation we obtain two more possibilities, see Fig. 5.15 (c), (d).

We assert that for each of those 4 possibilities either there is a mediator, or reductions along c_1, c_2 determine the same edge of Γ (this will complete the proof of Theorem 5.23).

Proofs are similar in all these cases, so here we only consider case (a). The boundary ∂N consists of 6 circles, and the complement $F \setminus \text{Int} N$ can consist of 4, 5, or 6 parts X_i, $1 \leq i \leq 6$, where the parts X_1, X_3, and also the parts X_5, X_6, may coincide. The reason is that only these two pairs lie on the same side of each of the circles c_1, and c_2, see Fig. 5.16.

Case 1. Suppose that $X_1 \neq X_3$. We are going to show that then the circles c_1, c_2 admit a mediator circle. Note that the circles $\partial X_1, \partial X_2$ and ∂X_3 are sep-

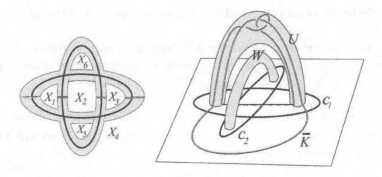

Figure 5.16

arating and intersect $\bar{K}$ at 2 points each. If neither of them is a mediator, then they bound discs with trivial arcs inside. This contradicts the non-triviality of the circle c_1, which encircles parts X_1, X_2 and X_3.

Case 2. Suppose that $X_1 = X_3$. Then they form a single surface, which we denote by U. The parts ∂X_5 and ∂X_6 may also form a single surface W, see Fig. 5.16 to the right.

In this situation reductions along c_1 resp. c_2 give identical results, (the prime summand corresponding to W being just a thickening of a torus with knot curve a meridian), so property (MF2) is automatically true.

Using the Diamond Lemma we may conclude that Theorem 5.23 is also true. □

We close by exhibiting an example of a homologically nontrivial knot in a thickened genus two surface F having two different prime decompositions.

Example 5.24 *Assume a thickened genus 2-surface F is given as in the upper part of Fig. 5.17. We assume that the projection $\bar{K}$ of K intersects two meridional annuli U, W of the handles and the disc T in a sufficiently complicated way such that the corresponding knots in U, W and T are nontrivial and different.*

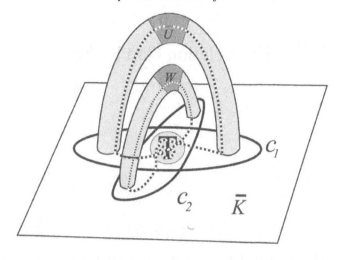

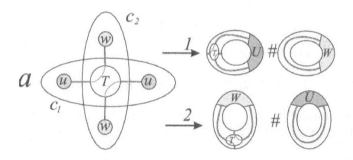

Figure 5.17 A knot having two different prime decompositions

Performing reductions of K along c_1 resp. c_2 produces different prime decompositions of K into two different pairs of knots in thickened tori, see the lower part of Fig. 5.17.

6

The Relation Gap Problem

Jens Harlander

6.1 Introduction

The augmentation ideal IG of a group can be thought of as a linear model of G. A presentation

$$N \to F(\mathbf{x}) \xrightarrow{\phi} G$$

of G gives rise to a $\mathbb{Z}G$-module presentation

$$N/[N,N] \xrightarrow{\partial} \mathbb{Z}G(\tilde{\mathbf{x}}) \xrightarrow{\partial_1} IG$$

of IG. In particular, if G has an n-generator m-relator presentation, then so does IG. Natural questions are

1. Can IG be generated by fewer elements than needed to generate G?
2. If a presentation of G requires m relators, can one have fewer than m relators in the corresponding presentation of IG?

Karl Gruenberg called the difference

$$d(G) - d_G(IG)$$

between the minimal numbers of generators of G and the minimal number of generators of the left $\mathbb{Z}G$-module IG the *generation gap* of G, and referred to the first question as the *generation gap problem*. He called the difference

$$d_F(N) - d_G(N/[N,N])$$

between the minimal number of normal generators of N and the the minimal number of generators of the left $\mathbb{Z}G$-module $N/[N,N]$ the *relation gap* of F/N, and referred to the second question as the *relation gap problem*.

The generation gap problem was solved in 1974 by Cossey, Gruenberg, and Kovacs. They showed that arbitrarily large generation gaps can occur for finite groups. The relation gap problem for finitely presented groups, however, is still open. The mathematics employed in search of an answer ranges from representation theory and homological group theory to techniques in geometric group theory and topology. Two results deserve special mention, since they present, in a sense, extreme cases of the problem.

The first says that for finite groups the rank of the relation module can be computed and the difference

$$d(F) - d_G(N/[N,N])$$

is a group invariant and does not depend on the choice of the finite presentation F/N. Thus, for finite groups, the relation gap question comes down to the question of whether the difference

$$d(F) - d_F(N)$$

is a group invariant.

The second result says that infinite relation gaps can occur for finitely generated infinite groups. This was shown by Bestvina and Brady in the early 1990's by developing geometric techniques such as Morse theory in the context of cubical complexes.

In this chapter we survey results and methods, both algebraic and geometric in nature, that arose from work on the relation gap problem.

6.2 The Relation Gap problem

Let F/N be a finite presentation of the group G. We have

$$d_F(N) \geq d_G(N/[N,N]) \geq d(N/[F,N]) \geq \dim_{\mathbf{F}}(\mathbf{F} \otimes (N/[F,N])).$$

Here $d_F(-)$ denotes the minimal number of F-group generators (since the action of F on N is conjugation, this is just the minimal number of normal generators of N), $d_G(-)$ denotes the minimal number of G-module generators, $d(-)$ denotes the minimal number of generators, and $\dim_{\mathbf{F}}(-)$, is the $\mathbf{F}$-vector space dimension, where $\mathbf{F}$ is a field.

Definition 6.1 *Let F/N be a finite presentation of a group G.*

1. *F/N has a* relation gap *if $d_F(N) > d_G(N/[N,N])$;*
2. *F/N is* efficient *if $d_F(N) = d(N/[F,N])$;*
3. *F/N is* proficient *if $d_G(N/[N,N]) = d(N/[F,N])$*

Note that an efficient presentation does not have a relation gap, and a proficient presentation that is not efficient does have a relation gap.

Relation gap problem: Does there exist a finite presentation F/N with a relation gap?

Infinite relation gaps are known to exist for finitely generated groups. See Bestvina and Brady [BeBr97]. More about that in a later section.

Denote by $r(A)$ the torsion-free rank of a finitely generated abelian group A.

Lemma 6.2 *Let F/N be a finite presentation of G. Then*

1 $d(N/[F, N]) = d(F) - r(H_1(G)) + d(H_2(G))$;
2 $\dim_\mathbf{F}(\mathbf{F} \otimes (N/[F, N])) = d(F) - \dim_\mathbf{F}(H_1(G, \mathbf{F})) + \dim_\mathbf{F}(H_2(G, \mathbf{F}))$;
3 $\dim_\mathbf{F}(\mathbf{F} \otimes (N/[F, N])) = d(F) - \dim_\mathbf{F}(H^1(G, \mathbf{F})) + \dim_\mathbf{F}(H^2(G, \mathbf{F}))$.

Proof: Recall that an exact sequence of groups

$$1 \to H \to G \to Q \to 1$$

gives rise to an exact sequence

$$H_2(G, M) \to H_2(Q, M) \to M \otimes_{\mathbb{Z}Q} H/[H, H] \to H_1(G, M) \to H_1(Q, M) \to 0,$$

for each $\mathbb{Z}G$-module M. See Brown [I], [Br82] for details. If we apply this to

$$1 \to N \to F \to G \to 1$$

we obtain

$$0 \to H_2(G, M) \to M \otimes_{\mathbb{Z}Q} N/[N, N] \to H_1(F, M) \to H_1(G, M) \to 0.$$

If we take $M = \mathbb{Z}$ with trivial G-action, then we obtain

$$0 \to H_2(G) \to N/[F, N] \to H_1(F) \to H_1(G) \to 0$$

and (1) follows. If we take $M = \mathbf{F}$ with trivial G action we obtain

$$0 \to H_2(G, \mathbf{F}) \to \mathbf{F} \otimes N/[F, N] \to H_1(F, \mathbf{F}) \to H_1(G, \mathbf{F}) \to 0$$

and (2) follows. Finally, (3) follows from (2) because $H_i(G, \mathbf{F})$ is isomorphic to $H^i(G, \mathbf{F})$, $i = 1, 2$, for a finitely presented group G. This follows from the universal coefficient theorem. □

Lemma 6.3 *Let F/N be a presentation of a group G. Let $\mathbf{s}$ be a subset of N and let J be the normal closure of $\mathbf{s}$ in F. Then the following statements are equivalent:*

1 The elements $s[N,N]$, $s \in \mathbf{s}$ generate the relation module $N/[N,N]$;
2 The subgroup N/J of F/J is perfect, that is $(N/J)_{ab} = 0$.

Proof: Note that $(N/J)_{ab} = N/J[N,N]$. Thus $(N/J)_{ab} = 0$ if and only if $N = J[N,N]$. The last equation holds if and only if the elements $s[N,N]$, $s \in \mathbf{s}$ generate the relation module $N/[N,N]$. This shows the equivalence of (1) and (2). □

Theorem 6.4 *Let F/N be a finite presentation of G and assume that $d_G(N/[N,N]) = m$. Let $s_1[N,N], ..., s_m[N,N]$ be a set of relation module generators and let J be the normal closure of $\mathbf{s} = \{s_1, ..., s_m\}$ in F. Let $\hat{G}$ be the group presented by F/J. Then*

1 F/J does not have a relation gap;
2 If F/N is proficient then F/J is efficient.

Proof: We have a $\mathbb{Z}\hat{G}$-epimorphism

$$J/[J,J] \to N/[N,N]$$

and hence also a group epimorphism

$$J/[F,J] \to N/[F,N].$$

Thus

$$d_{\hat{G}}(J/[J,J]) \geq d_G(N/[N,N]) \tag{6.1}$$

and

$$d(J/[F,J]) \geq d(N/[F,N]). \tag{6.2}$$

By the way J was constructed we also have

$$d(J/[F,J]) \leq d_{\hat{G}}(J/[J,J]) \leq d_F(J) \leq d_G(N/[N,N]). \tag{6.3}$$

Inequalities (1) and (3) imply that $d_F(J) = d_{\hat{G}}(J/[J,J])$, so F/J does not have a relation gap. Suppose that F/N is proficient. Then

$$d(J/[F,J]) \leq d_{\hat{G}}(J/[J,J]) \leq d_F(J) \leq d_G(N/[N,N]) = d(N/[F,N]). \tag{6.4}$$

Now inequalities (2) and (4) imply that all $\leq$ in (4) are =. Hence $d_F(J) = d(J/[F, J])$ and so F/J is efficient. □

A geometric view of the relation module was developed in the first chapter of this book. We end this section by recalling the main result. Given a presentation

$$N \to F(\mathbf{x}) \xrightarrow{\phi} G$$

of a group G, we have a 4-term exact sequence

$$0 \to N/[N, N] \xrightarrow{\partial} C_1(\Gamma(G, \mathbf{g})) \xrightarrow{\partial_1} C_0(\Gamma(G, \mathbf{g})) \xrightarrow{\epsilon} \mathbb{Z} \to 0. \qquad (6.5)$$

Here $\mathbf{g} = \phi(\mathbf{x})$, and $\Gamma(G, \mathbf{g})$ is the Cayley graph of G associated with the generating set $\mathbf{g}$, and $C_i(\Gamma(G, \mathbf{g}))$ is the ith-chain group of the Cayley graph. The map ∂ is defined in terms of the Fox derivatives,

$$\partial(r[N, N]) = \sum_{x \in \mathbf{x}} \phi(\frac{\partial r}{\partial x}) \tilde{x}.$$

The 4-term exact sequence shows that the relation module is isomorphic to the first homology of the Cayley graph.

6.3 Examples

Example 1: Consider the presentation

$$P = \langle a, b, c, d \mid [a, b], a^m, [c, d], c^n \rangle$$

of the free product $G = (\mathbb{Z}_m \oplus \mathbb{Z}) * (\mathbb{Z}_n \oplus \mathbb{Z})$, where m and n are relatively prime. Let F be the free group on the generators of P and let N be the normal closure or the relators in P. Note that $d(N/[F, N]) = 4 - 2 + 1 = 3$ by Lemma 6.2. Epstein [Eps61] asked if this presentation is efficient. Gruenberg and Linnell [GrLi08] showed that $d_G(N/[N, N]) = 3$. Questions concerning presentations of free products are discussed in detail in Chapter 7.

Example 2: The following examples were constructed by Bridson and Tweedale [BrTw07]. They are very much in the spirit of the Epstein example above, but an unexpectedly small set of relation module generators can be seen more easily. More details on these examples can also be found in Chapter 7 of this book. Let G_n be the group defined by

$$P_n = \langle a, b, x \mid a^n, b^n, [a, b], xax^{-1}b^{-1} \rangle.$$

This group is an HNN-extension of $\mathbb{Z}_n \times \mathbb{Z}_n$ where the stable letter x conjugates one factor into the other. Note that

$$P'_n = \langle a, x \mid a^n, [a, xax^{-1}] \rangle$$

also presents G_n. Let $\rho_n(a, x) = [xax^{-1}, a]a^{-n}$. Then

$$P'_m * P'_n = \langle a, x, b, y \mid a^m, [a, xax^{-1}], b^n, [b, yby^{-1}] \rangle$$

is a presentation of the free product $G_m * G_n$. Let F be the free group on the generators and let N be the normal closure of the relators in $P'_m * P'_n$. Bridson and Tweedale showed that the relation module N_{ab} is generated by $\rho_m(a, x)[N, N]$, $\rho_n(b, y)[N, N]$, and $a^m b^{-n}[N, N]$.

Another article by Bridson and Tweedale that addresses the relation gap is [BrTw14].

Example 3: The following construction first appeared in [Har00]. Let F_1/N_1 and F_2/N_2 be finite presentations of groups G_1 and G_2, respectively. Let H be a finitely generated subgroup of both G_1 and G_2 and let F/N be the standard presentation of the amalgamated product $G = G_1 *_H G_2$ obtained from the presentations F_i/N_i, $i = 1, 2$, and a finite generating set of H. One can show that

$$d_G(N_{ab}) \leq d_{G_1}(N_{1ab}) + d_{G_2}(N_{2ab}) + d_H(IH),$$

where IH is the augmentation ideal of H. Denote by H^n the n-fold direct product $H \times \ldots \times H$. Cossey, Gruenberg, and Kovacs [CoGrKo74] showed that $d_{H^n}(IH^n) = d_H(IH)$ in case H is a finite perfect group. Since $d(H^n) \to \infty$ as $n \to \infty$ one can produce arbitrarily large generation gaps

$$d(H^n) - d_{H^n}(IH^n).$$

This leads to unexpectedly small generating sets for the relation module N_{ab} for presentations F/N of $G_1 *_{H^n} G_1$. The hope is that the amalgamated product "shifts" the generation gap into a relation gap.

6.4 An Infinite Relation Gap

We outline the work of Bestvina and Brady [BeBr97] on finiteness conditions of groups. This section is an abbreviated version Section 8.3 of Ross Geoghegan's book on topological methods in group theory [Geo08].

Recall that a group G is of *type F_n* if there is a $K(G, 1)$ complex with finite n-skeleton. Type F_1 is equivalent to the group being finitely generated, and type F_2 is equivalent to the group being finitely presentable. The group G is of *type FP_n* if there is a projective resolution

$$\cdots \to P_{n+1} \to P_n \to \cdots \to P_0 \to \mathbb{Z} \to 0$$

where the P_i are projective $\mathbb{Z}G$-modules that are finitely generated for $0 \le i \le n$. It is known that a group is of type FP_1 if and only if it is of type F_1, and type FP_2 is equivalent to the statement that G has a presentation F/N, where F is finitely generated and the relation module $N/[N, N]$ is finitely generated. Thus, if G is a group of type FP_2 but not of type F_2, then any presentation F/N, where F is finitely generated, has an infinite relation gap: $d_G(N/[N, N]) < d_F(N) = \infty$.

Let A and B be simplicial complexes with vertex sets V_A and V_B, respectively. The *join* $A * B$ is a simplicial complex defined as follows: its vertex set is $V_A \cup V_B$, and $\sigma \subseteq V_A \cup V_B$ is a simplex of $A * B$ if $\sigma \cap V_A$ is a simplex of A and $\sigma \cap V_B$ is a simplex of B.

A *flag complex* L is a simplicial complex that is determined by its 1-skeleton: a finite set of vertices of L span a simplex in L if and only if each pair of those vertices span a 1-simplex. A finite flag complex determines a group presentation

$$P = \langle \mathbf{s} \mid \mathbf{r} \rangle,$$

where $\mathbf{s}$ is the set of vertices of L and

$$\mathbf{r} = \{[s_i, s_j] = 1 \mid s_i \text{ and } s_j \text{ span a 1-simplex of } L\}.$$

The presentation defines a *right angled Artin group* G. Let T be the product

$$T = \prod_{s \in L^{(0)}} S_s^1$$

of circles S_s^1, each carrying the standard CW-structure consisting of one vertex and one edge. The torus T is a CW-complex with a single vertex u and the link at u is simplicially isomorphic to a join of 0-spheres:

$$\mathrm{lk}(T, u) = *_{s \in S} S_s^0.$$

Each simplex σ of L defines a subtorus $T_\sigma = \prod_{s \in \sigma} S_s^1$ of T. Define $Z = \bigcup_{\sigma \in L} T_\sigma$. We have $\mathrm{lk}(T_\sigma, u) = *_{s \in \sigma} S_s^0$ and $\mathrm{lk}(Z, u) = \bigcup_{\sigma \in L} \mathrm{lk}(T_\sigma, u)$. It follows that $\mathrm{lk}(Z, u)$ contains copies of L, and in fact, this link contains L as a retract. Note that $\pi_1(Z) = G$. The universal covering X of Z is a cubical complex. We give each cube in X the metric of the standard unit cube and give X

the associated path metric. One can show than this makes X into a CAT(0)-metric space. In particular X is contractible and hence Z is a $K(G, 1)$-complex. Let $f_0 \colon Z \to S^1$ be the map that takes every 1-cell of Z homeomorphically to the circle S^1 so that the induced map $\phi \colon G \to \mathbb{Z}$ maps every generator $s \in \mathbf{s}$ to 1. Let $f \colon X \to \mathbb{R}$ be the lift of f_0. We think of f as a height function on X. If $t \in \mathbb{R}$ we define the level set $X_t = f^{-1}(t)$. More generally, if $J \subset \mathbb{R}$ is a closed subset, we define $X_J = f^{-1}(J)$. Let v be a vertex in X. The *ascending link* $\mathrm{lk}^{\uparrow}(X, v)$ is the subcomplex of $\mathrm{lk}(X, v)$ spanned by the vertices of $\mathrm{lk}(X, v)$ that come from edges e of X that contain v and $f|_e$ has a minimum at v. The *descending link* $\mathrm{lk}^{\downarrow}(X, v)$ is defined similarly, replacing "minimum" by "maximum". Using the description of $\mathrm{lk}(Z, u)$ given above one can check that both $\mathrm{lk}^{\uparrow}(X, v)$ and $\mathrm{lk}^{\downarrow}(X, v)$ are simplicially isomorphic to L, and both are retracts of $\mathrm{lk}(X, v)$. Another crucial observation is that the ascending and the descending links control the connectedness properties of the level sets.

Lemma 6.5 *Let $J \subseteq J'$ be closed connected subsets of $\mathbb{R}$ such that $\inf J = \inf J'$ and $J' - J$ contains only one point t of $f(X^{(0)})$. Then*

$$X_J \cup \bigcup \{\text{cone on } \mathrm{lk}^{\downarrow}(X, v) \mid f(v) = t\}$$

is a strong deformation retract of $X_{J'}$.

For a proof see Geoghegan [Geo08], Proposition 8.3.3.

Theorem 6.6 *Let L be a finite flag complex and let G be the corresponding right-angled Artin group. Let $\phi \colon G \to \mathbb{Z}$ be the epimorphism taking all generators to $1 \in \mathbb{Z}$, and let $H = \ker(\phi)$. Then*

1 H has type F_n if and only if L is $(n-1)$-connected;
2 H has type FP_n if and only if L is $(n-1)$-acyclic.

Corollary 6.7 *Let L be a connected finite flag complex and assume that $H_1(L) = 0$, but $\pi_1(L) \neq 1$. Let H be as in Theorem 6.6. Then H is of type FP_2 but not of type F_2. Consequently, any presentation F/N of H, where F is a finitely generated, has an infinite relation gap.*

Proof: We sketch a proof of Theorem 6.6. Assume first that L is $(n-1)$-connected. Consider the level set X_0. Let $\alpha \in \pi_i(X_0)$, $1 \leq i \leq n-1$. Since X is contractible α the trivial element of $\pi_i(X_{[-k,k]})$ for some k. Since by Lemma 6.5 the space $X_{[-k,k]}$ is obtained from X_0 by repeatedly coning off $(n-1)$-connected subspaces, it follows that the inclusion induced map $\pi_i(X_0) \to \pi_i(X_{[-k,k]})$ is an isomorphism for $1 \leq i \leq n-1$. Thus α is trivial in $\pi_i(X_0)$. It follows that X_0 is $(n-1)$-connected. Since H acts freely and co-compactly on X_0, it follows that H is of type F_n.

For the other direction consider the filtration $\{X_{[-k,k]} \mid k \in \mathbb{N}_0\}$ of X. Note that the group H acts freely and co-compactly on each $X_{[-k,k]}$. Thus, according to Brown's criterion (see Brown [Bro87] or Geoghegan [Geo08], Theorem 7.4.1) H is of type F_n if and only if the filtration is essentially $(n-1)$-connected. That means for every k_1 there exists $k_2 \geq k_1$ so that the inclusion induced map $\pi_i(X_{[-k_1,k_1]}) \to \pi_i(X_{[-k_2,k_2]})$ is trivial, for all $0 \leq i \leq n-1$. Let us assume that L is not $(n-1)$-connected and $\pi_j(L) \neq 0$ for some $0 \leq j \leq n-1$. We will show that $\pi_j(X_0) \to \pi_j(X_{[-k,k]})$ is not trivial for any k. Given a k choose a vertex v in X so that $f(v) > k$. There is a retraction $r: X - \{v\} \to \text{lk}(X,v)$ defined as follows: for $x \in X - \{v\}$ let $[v,x)$ be the geodesic ray in X that starts at v and passes through x. Let $r(x) = [v,x) \cap \text{lk}(X,v)$. We also have a map $g: \text{lk}^\downarrow(X,v) \to X_0$ defined by $g(x) = [v,x) \cap X_0$. Let $\alpha: S^j \to \text{lk}^\downarrow(X,v) \subseteq X - \{v\}$ be a map that represents a nontrivial element of $\pi_j(\text{lk}^\downarrow(X,v))$. Note that since $\text{lk}^\downarrow(X,v)$ is a retract of $\text{lk}(X,v)$, this implies that α does not represent the trivial element in $\pi_j(\text{lk}(X,v))$. Let $\beta = g \circ \alpha: S^j \to X_0$. If β represents the trivial element of $\pi_j(X_{[-k,k]})$, then it also represents the trivial element of $\pi_j(X - \{v\})$ because $X_{[-k,k]} \subseteq X - \{v\}$. But then $r \circ \beta = \alpha$ represents the trivial element of $\pi_j(\text{lk}(X,v))$, which is not the case. This shows that $\pi_j(X_0) \to \pi_j(X_{[-k,k]})$ is not trivial, and hence the filtration $\{X_{[-k,k]} \mid k \in \mathbb{N}_0\}$ is not essentially $(n-1)$-connected. This proves the first statement in Theorem 6.6.

The homotopical arguments in the proof just given can be replaced by analogous homological arguments. See [Geo08], proof of Theorem 8.3.12. □

Another remarkable result contained in [BeBr97] states that either Whitehead's asphericity conjecture or the Eilenberg-Ganea conjecture is false. Chapter X in [I] and Chapter 4 in this book are concerned with the first conjecture. The latter conjecture was mentioned in Chapter 1.

Whitehead's Asphericity Conjecture: A subcomplex of an aspherical 2-complex is aspherical.

Eilenberg-Ganea Conjecture: A group of cohomological dimension 2 has geometric dimension 2.

Let L be a flag complex that arises as a suitable subdivision of the standard 2-complex built from the presentation

$$P = \langle x, y, z \mid x^2 = y^3 = z^5 = xyz \rangle$$

of the binary icosahedral group. Direct computation shows that L is acyclic:

$H_i(L) = 0$ for all $i \geq 1$. Let G be the corresponding right-angled Artin group. Let $\phi\colon G \to \mathbb{Z}$ be the epimorphism taking all generators to $1 \in \mathbb{Z}$, and let $H = \ker(\phi)$.

Theorem 6.8 *Either H is a counterexample to the Eilenberg-Ganea conjecture, or there is a counterexample to the Whitehead conjecture.*

We sketch the argument given in [BeBr97]. A different proof was given by Howie in [How99].

Proof: Note that X is 3-dimensional and hence the level set X_0 is 2-dimensional. Since L is acyclic X_0 is acyclic as well, and since H acts freely and co-compactly on X_0 it follows that H has cohomological dimension 2. If H does not have geometric dimension 2 then the Eilenberg-Ganea conjecture is false. Suppose H does have geometric dimension 2. Then there exists a contractible 2-complex Y on which H acts freely. Since H is FP_∞ by Theorem 6.6 it is finitely generated and we may assume H acts co-compactly on the 1-skeleton of Y. Choose an H-equivariant quasi-isometry

$$q\colon X_0 \to Y.$$

Preimages of points of a quasi-isometry have bounded diameter, say $q^{-1}(y) \leq C$. For any $v \in X$ with height $f(v) > 0$ define $S(L, v) = g(\mathrm{lk}^\downarrow(X, v)) \subseteq X_0$. The map $g\colon \mathrm{lk}^\downarrow(X, v) \to X_0$ was define in the proof of Theorem 6.6. Note that the bigger we make $f(v)$, the "larger" the subcomplex $S(L, v) \subseteq X_0$ becomes. Let $q_v\colon S(L, v) \to q(S(L, v))$. One can show that if $f(v)$ is very large in comparison with the constant C, preimages of simplices in $q(S(L, v))$ are contained in contractible subsets of $S(L, v)$. Thus, for $f(v)$ large enough, one can define a homotopy inverse $\bar{q}_v\colon q_v(S(L, v)) \to S(L, v)$ to q_v. In particular the map $\pi_2(S(L, v)) \to \pi_2(q(S(L, v)))$ induced by q_v is injective. Since the fundamental group L is non-trivial and finite, $\pi_2(L) \neq 0$. Since $S(L, v)$ is homeomorphic to L, we have $\pi_2(S(L, v)) \neq 0$ and hence $\pi_2(q(S(L, v))) \neq 0$. Thus $q(S(L, v))$ is a subcomplex of the contractible complex Y that is not aspherical. □

6.5 Efficiency

Efficiency and proficiency of a group presentation F/N were explained in Definition 6.1.

Theorem 6.9 *A finite presentation F/N of a finitely generated abelian group is efficient. A finite presentation F/N of a finite nilpotent group is proficient.*

Proof: The first statement is elementary, a proof can be found in Harlander [I], [Ha92]. The second result is much deeper and is due to Wamsley [Wam70]. The idea of the proof is given in this chapter in the section on finite groups. □

We next discuss results of Lustig on minimality and efficiency of group presentations. We begin with a topological view of efficiency. Let G be a finitely presented group and K be a finite $(G, 2)$-complex. Note that the Euler characteristic of K is bounded below by a bound that depends only on the fundamental group. Indeed,

$$\chi(K) = \dim_{\mathbf{F}} H_0(K, \mathbf{F}) - \dim_{\mathbf{F}} H_1(K, \mathbf{F}) + \dim_{\mathbf{F}} H_2(K, \mathbf{F})$$
$$\leq 1 - r(H_1(G)) + d(H_2(G))$$

where $\mathbf{F}$ is a field. Recall from Chapter 1 that

$$\chi_{geom}(G, 2) = \min\{\chi(K) \mid K \text{ is a finite } (G, 2)\text{-complex}\}.$$

We say the K is *minimal* if $\chi(K) = \chi_{geom}(G, 2)$, and we say that K is *efficient* if $\chi(K) = 1 - r(H_1(G)) + d(H_2(G))$. If

$$P = \langle x_1, ..., x_n \mid r_1, ..., r_m \rangle$$

is a presentation of G then we say P is *minimal*, or *efficient*, if the associated standard 2-complex $K(P)$ is minimal, or efficient, respectively.

Lemma 6.10 *Let $F(x_1, ..., x_n)/N$ be a finite presentation of the group G. Assume that $d_F(N) = m$, and $\{r_1, ...r_m\}$ is a set of normal generators of N. Let*

$$P = \langle x_1, ..., x_n \mid r_1, ..., r_m \rangle.$$

The following statements are equivalent:

1 $F(x_1, ..., x_n)/N$ is efficient;
2 P is efficient.

Proof: Left to the reader. □

We say a group is *efficient* if it admits an efficient presentation. Non-efficient finite groups were first constructed by Swan (see [I], [Sw65]). Let $H = (\mathbb{Z}_7)^n$. Then $\mathbb{Z}_3$ acts on H by squaring, and we can form the semi-direct G of H with

$\mathbb{Z}_3$. Swan shows that G is not efficient. More details on Swan's construction can also be found in Harlander [Har00].

Non-efficient infinite groups were first constructed by Lustig [I], [Lu93], [Lus95]. In order to explain some of his ideas we need to define the *Fox ideals*. Let K be a (G,m)-complex, that is a finite connected m-dimensional complex so that $\pi_1(K) = G$ and $\pi_i(K) = 0$ for $2 \le i \le m-1$. Let $(C_*(\tilde{K}), \partial)$ be the cellular chain complex of the universal covering $\tilde{K}$ of K. Let $\mathbf{c} = \{c_1, ..., c_k\}$ be the set of j-cells in K and let $\tilde{\mathbf{c}} = \{\tilde{c}_1, ..., \tilde{c}_k\}$ be a set of lifts. The set $\tilde{\mathbf{c}}$ is a basis of the free $\mathbb{Z}G$-module $C_j(\tilde{K})$. Every element in the kernel of $\partial_j \colon C_j(\tilde{K}) \to C_{j-1}(\tilde{K})$ can be uniquely expressed as a linear combination $\alpha_1 \tilde{c}_1 + ... + \alpha_k \tilde{c}_k$, $\alpha_i \in \mathbb{Z}G$. Define $I_j(K)$ to be the 2-sided ideal in $\mathbb{Z}G$ generated by elements α that occur as coefficients of elements of $\ker(\partial_j)$.

Let R be a ring with $1 \ne 0$ that has the rank invariance property properties, that is $d_R(R^n) = n$. A ring of matrices over a commutative ring is a good choice for R.

Theorem 6.11 *(Lustig [I], [Lu93]) Let*

$$P = \langle x_1, ..., x_n \mid r_1, ..., r_m \rangle$$

be a presentation of a group G and assume there is a ring homomorphism $\rho \colon \mathbb{Z}G \to R$ such that $\rho(1) = 1$ and $\rho(I_2(K(P))) = 0$. Then P is minimal.

Lustig uses his result to construct minimal non-efficient presentations. Here are some details. Let H be the figure-8 knot group and let $G = H \times \mathbb{Z}$. We have group epimorphisms

$$G \to H \to H_{ab}$$

which induce a ring homomorphism

$$\psi \colon \mathbb{Z}G \to \mathbb{Z}H_{ab}.$$

Since H_{ab} is infinite cyclic, generated by t, say, the group ring $\mathbb{Z}H_{ab}$ is a Laurent-polynomial ring $\mathbb{Z}[t, t^{-1}]$. Let $\Delta(\mathbf{k}) \in \mathbb{Z}[t, t^{-1}]$ be the Alexander polynomial of the figure-8 knot $\mathbf{k}$, and let Δ be the ideal generated by the Alexander polynomial. Since $\Delta \ne \mathbb{Z}[t, t^{-1}]$, the ring $R = \mathbb{Z}[t, t^{-1}]/\Delta$ is a non-trivial ring. Let

$$\rho \colon \mathbb{Z}G \to \mathbb{Z}[t, t^{-1}]/\Delta = R.$$

The figure-8 knot group H can be presented with two generators and one relator, so G has a presentation

$$P = \langle a, b, c \mid r, [a, c], [b, c] \rangle.$$

One can compute $I_2(K(P)) \subseteq \mathbb{Z}G$ and check that the image of that ideal in $\mathbb{Z}[t, t^{-1}]$ is Δ. Thus

$$\rho(I_2(K(P))) = 0,$$

and the conditions of Theorem 6.11 are satisfied. It follows that P is minimal. The presentation P however is not efficient. Indeed, since $H_1(G) = \mathbb{Z} \times \mathbb{Z}$ and $H_2(G) = \mathbb{Z}$, we have

$$\chi(K(P)) = 1 - 3 + 3 = 1 > 1 - r(H_1(G)) + d(H_2(G)) = 1 - 2 + 1 = 0.$$

Since P is minimal and not efficient, it follows that G does not have an efficient presentation.

Minimal presentations that are not efficient are good candidates for having a relation gap. The next result says that if minimality of a presentation P is detected using Lustig's test, then P does not have a relation gap.

Theorem 6.12 *Let $P = \langle x_1, ..., x_n \mid r_1, ..., r_m \rangle$ be a presentation of a group G. Let F be the free group on $x_1, ..., x_n$ and N be the normal subgroup generated by $r_1, ..., r_m$. Assume there is a ring homomorphism $\rho \colon \mathbb{Z}G \to R$ such that $\rho(1) = 1$ and $\rho(I_2(K(P))) = 0$. Then $d_F(N) = d_G(N/[N, N]) = m$. In particular F/N does not have a relation gap.*

Proof: Consider the chain complex

$$C_2(\tilde{K}(P)) \xrightarrow{\partial_2} C_1(\tilde{K}(P)) \xrightarrow{\partial_1} C_0(\tilde{K}(P)) \xrightarrow{\epsilon} \mathbb{Z} \to 0$$

of the universal covering $\tilde{K}(P)$ of $K(P)$. The 4-term exact sequence given at the end of Section 6.2 shows that $C_2(\tilde{K}(P))/\ker(\partial_2)$ is isomorphic to the relation module $N/[N, N]$. Let $\{c_1, ..., c_n\}$ be the 2-cells in $K(P)$. Then $C_2(\tilde{K}(P))$ has a $\mathbb{Z}G$-basis $\{\tilde{c}_1, ..., \tilde{c}_m\}$, where $\tilde{c}_i$ is a lift of c_i. We have a surjection

$$C_2(\tilde{K}(P)) = \mathbb{Z}G^m \to R^m$$

defined by

$$\sum_{i=1}^{m} \alpha_i \tilde{c}_i \to (\rho(\alpha_1), ..., \rho(\alpha_m)),$$

which factors through $\ker(\partial_2)$ because $\rho(I_2(K(P))) = 0$.

Thus
$$d_G(C_2(\tilde{K}(P))/\ker(\partial_2)) = d_G(N/[N,N]) \geq d_R(R^m) = m.$$

□

Note that a group G as in Theorem 6.12 might still have a non-minimal presentation that has a relation gap.

In the remaining of this section we will discuss a result concerning embeddings into efficient groups. At the heart of the proof is a a stabilization "trick" due to Hog-Angeloni and Metzler which was used in work on the homotopy and simple homotopy types of 2-complexes [I], [Ho-AnMe90], [I], [Me90]. In [Har96] a generalized version of the trick is discussed and used to show that a relation gap can be closed via stabilization. This is strengthened in [Har97] which contains the following theorem.

Theorem 6.13 *Let G be a finitely presented group. Then there exists $k \geq 1$ such that*

$$G * (\mathbb{Z}_2 \times \mathbb{Z}_2) * \ldots * (\mathbb{Z}_2 \times \mathbb{Z}_2)$$

(k free factors of $(\mathbb{Z}_2 \times \mathbb{Z}_2)$) is efficient. There also exists a number $l \geq 1$ such that

$$G \times (\mathbb{Z}_2 \times \mathbb{Z}_2) \times \ldots \times (\mathbb{Z}_2 \times \mathbb{Z}_2)$$

(l direct factors of $(\mathbb{Z}_2 \times \mathbb{Z}_2)$) is efficient. In particular, a finite group embeds into a finite efficient group.

Here is the main idea of the proof. Let $F(\mathbf{x})/N$ be a finite presentation of G, where $\mathbf{x} = \{x_1, \ldots, x_n\}$. Suppose the relation module is generated by $s_1[N, N], \ldots, s_m[N, N]$. Then N is normally generated by $s_1, \ldots, s_m$ together with a finite set of elements of the form $[t, t']$, where $t, t' \in N$. For simplicity assume that $N = \langle\langle s_1, \ldots, s_m, [t, t']\rangle\rangle$. Now note that

$$P_1 = \langle x_1, \ldots x_n, a, b \mid s_1, \ldots, s_m, [t, t'], a^2, b^2, [a, b]\rangle$$

and

$$P_2 = \langle x_1, \ldots x_n, a, b \mid s_1, \ldots, s_m, ta^{-2}, t'b^{-2}, [a, b]\rangle$$

both present the group $G * (\mathbb{Z}_2 \times \mathbb{Z}_2)$. That P_1 presents the free product is clear. Let us take a look at P_2. In the group $\bar{G}$ presented by P_2 the equations $t = a^2$ and $t' = b^2$ hold. Since a and b commute in $\bar{G}$, the squares a^2 and b^2 also commute, hence t and t' commute in $\bar{G}$. So the equation $[t, t'] = 1$ holds. Since $N = \langle\langle s_1, ..., s_m, [t, t']\rangle\rangle$, we get $t = 1$ and $t' = 1$ in $\bar{G}$, and hence $a^2 = 1$ and $b^2 = 1$ in $\bar{G}$. Thus $\bar{G} = G * (\mathbb{Z}_2 \times \mathbb{Z}_2)$. The commutator of relators $[t, t']$ got "absorbed" in P_2, so this second presentation makes due with one less relator than the first presentation.

6.6 Presentations with Cyclic Relation Modules

Let F/N be a finite presentation of a group G and assume that

$$d_G(N/[N, N]) = 1.$$

Does it follow that $d_F(N) = 1$? This question was answered affirmatively under the additional assumption that G is solvable. See [I], [Ha92]. The question is still open in general.

Theorem 6.14 *Let F/N be a finite presentation of a solvable group G. If $d_G(N/[N, N]) = 1$ then $d_F(N) = 1$. Only the following two cases can occur:*

1. *F is generated by a single element x and we have $N = \langle\langle x^k \rangle\rangle$ for some $k \in \mathbb{Z}$;*
2. *F is generated by two elements x, y and we have $N = \langle\langle xyx^{-1}y^{-k} \rangle\rangle$, for some $k \in \mathbb{Z}$.*

The proof relies on Strebel's theory of E-groups. Recall that a group G is an *E-group* if G_{ab} is torsion-free and there exists a projective resolution

$$... \to P_2 \xrightarrow{\partial_2} P_1 \xrightarrow{\partial_1} P_0 \xrightarrow{\partial_0} \mathbb{Z} \to 0$$

so that

$$\mathbb{Z} \otimes_G \partial_2 : \mathbb{Z} \otimes_G P_2 \to \mathbb{Z} \otimes_G P_1$$

is injective. In particular $H_2(G) = 0$. In [I], [St74] Strebel showed that If G is an E-group, then so is the derived group $G^{(\alpha)}$ for every ordinal α. Furthermore, if $G^{(\alpha)} = G^{(\alpha+1)}$ (i.e. $G^{(\alpha)}$ is perfect), then $\alpha = 0, 1, 2$ or is a limit ordinal.

Let us sketch the proof of Theorem 6.14. Suppose that $N/[N, N]$ is generated by $s[N, N]$. Let J be the normal closure of s in F. Let $H = N/J$ and $\hat{G} = F/J$. By Lemma 6.3 the group H is perfect, and hence $H \leq \hat{G}^{(\alpha)}$ for every ordinal α. Since $\hat{G}/H$ is solvable, it follows that $H = \hat{G}^{(\alpha)}$ for some positive integer α. If $\alpha = 1$ the F/N presents an abelian group and hence is an efficient presentation by Theorem 6.9. In that case we are done. Assume $\alpha \geq 2$. Then $\hat{G}/\hat{G}^{(2)}$ is a metabelian image of a the finitely presented solvable group $G = \hat{G}/H$, and hence is finitely presented (see Bieri and Strebel [BiSt80]). It follows from a result of Baumslag [Bau74] that $d(F) \leq 2$ and that the 1-relator group $\hat{G}$ is torsion-free. Furthermore, since $\hat{G}$ is a 1-relator group and $\hat{G}/\hat{G}^{(2)}$ is finitely presented, the commutator subgroup $\hat{G}^{(1)}$ is an E-group (see Strebel [I], [St74]; see also [I], [Ha92] for a different argument). It follows that $H = \hat{G}^{(\alpha)}$ is an E-group, in particular $H_2(H) = 0$. Since $\hat{G}$ is a torsion-free one relator group, we have an exact sequence

$$0 \to \mathbb{Z}\hat{G} \xrightarrow{\partial_2} \mathbb{Z}\hat{G}^2 \xrightarrow{\partial_1} \mathbb{Z}\hat{G} \xrightarrow{\epsilon} \mathbb{Z} \to 0.$$

If we apply $\mathbb{Z} \otimes_H -$ we obtain an exact sequence

$$0 \to \mathbb{Z}G \xrightarrow{\partial_2} \mathbb{Z}G^2 \xrightarrow{\partial_1} \mathbb{Z}G \xrightarrow{\epsilon} \mathbb{Z} \to 0.$$

since $H_1(H) = H_2(H) = 0$. Thus G has cohomological dimension 2. The result follows from the classification of solvable groups of cohomological dimension 2 (see Gildenhuys [Gil76])

6.7 Finite Groups

The standard reference for this section is Gruenberg's book "Relation Modules of Finite Groups" [Gru76]. If x is a real number we denote by $\lceil x \rceil$, the ceiling of x, the smallest integer $\geq x$.

Theorem 6.15 *Let G be a finite group, F/N be a finite presentation of G. Then, for any field $\mathbf{F}$, $d_{\mathbf{F}G}(\mathbf{F} \otimes (N/[N,N]))$ is equal to*

$$\max_{M \in \mathcal{M}} \{\lceil \frac{\dim_{\mathbf{F}} H^2(G, M) - \dim_{\mathbf{F}} H^1(G, M) + \dim_{\mathbf{F}} H^0(G, M)}{\dim_{\mathbf{F}} M} \rceil + d(F) - 1\},$$

where $\mathcal{M}$ is the set of irreducible $\mathbf{F}G$-modules.

Note that for $M \in \mathcal{M}$ we have

$$H^0(G, M) = M^G = \begin{cases} \mathbf{F} \text{ if } M = \mathbf{F} \\ 0 \text{ if } M \neq \mathbf{F} \end{cases}.$$

We sketch a proof of Theorem 6.15. See [Gru76] for details. Let $J = J(\mathbf{F}G)$ be the Jacobson ideal of the group algebra $\mathbf{F}G$. So J is the intersection of all maximal left ideals. One can show that it annihilates every irreducible module. That is, if M is an irreducible $\mathbf{F}G$-module, then $JM = 0$. The quotient $\mathbf{F}G/J$ is semi-simple, and hence every $\mathbf{F}G/J$-module is a direct sum of irreducible $\mathbf{F}G/J$-modules. If V is an $\mathbf{F}G$-module and M is an irreducible $\mathbf{F}G$-module, define $e(V, M)$ to be the number of times M occurs in V/JV.

Lemma 6.16 $d_{\mathbf{F}G}(V) = \max\{\lceil \frac{e(V,M)}{e(\mathbf{F}G,M)} \rceil \mid M \text{ is an irreducible } \mathbf{F}G\text{-module}\}$

Proof: Assume first that $J = 0$: $\mathbf{F}G$ is semi-simple. In that case every module is the direct sum of irreducible modules and every submodule is a direct factor. Let $\{M_1, ..., M_l\}$ be the set of irreducible $\mathbf{F}G$-modules. For simplicity we write $e(V, M_i) = a_i$ and $e(\mathbf{F}G, M_i) = b_i$. Then

$$V = M_1^{a_1} \oplus \cdots \oplus M_l^{a_l}$$

and

$$\mathbf{F}G = M_1^{b_1} \oplus \cdots \oplus M_l^{b_l}.$$

Suppose that $d_{\mathbf{F}G}(V) = n$. Then $\mathbf{F}G^n$ maps onto V, and hence $\mathbf{F}G^n = V \oplus V'$. Thus

$$\mathbf{F}G^n = M_1^{nb_1} \oplus \cdots \oplus M_l^{nb_l} = M_1^{a_1} \oplus \cdots \oplus M_l^{a_l} \oplus V'.$$

It follows from the Krull-Schmidt theorem (see [Rot02]) that $nb_i \geq a_i$ for $i = 1, ..., l$. Let n' be another integer that satisfies $n'b_i \geq a_i$ for $i = 1, ..., l$. Then

$$\mathbf{F}G^{n'} = M_1^{n'b_1} \oplus \cdots \oplus M_l^{n'b_l}$$

contains

$$V = M_1^{a_1} \oplus \cdots \oplus M_l^{a_l}$$

as a submodule, and hence as a direct summand. Thus $\mathbf{F}G^{n'}$ maps onto V and hence $n' \geq n$. It follows that n is the smallest integer greater or equal to $\frac{a_i}{b_i}$, for all $i = 1, ..., l$.

Now assume $\mathbf{F}G$ is not semi-simple. In that case the arguments just given show the equation

$$d_{\mathbf{F}G/J}(V/JV) = \max\{\lceil \frac{e(V, M)}{e(\mathbf{F}G, M)} \rceil \mid M \text{ is an irreducible } \mathbf{F}G\text{-module}\}.$$

By Nakayama's lemma (see 3.14 in [Gru76]), $d_{\mathbf{F}G}(V) = d_{\mathbf{F}G/J}(V/JV)$. This completes the proof. $\square$

In the following we write $\bar{N} = \mathbf{F} \otimes (N/[N, N])$. In order to prove Theorem 6.15 we need to compute $e(\bar{N}, M)$. Let

$$\cdots P_3 \xrightarrow{d_3} P_2 \xrightarrow{d_2} P_1 \xrightarrow{d_1} P_0 \xrightarrow{\epsilon} \mathbf{F} \to 0$$

be a minimal projective resolution of $\mathbf{F}$ so that $\ker \epsilon = JP_0$ and $\ker d_k \subseteq JP_k$ for $k \in \mathbb{N}$. Let M be an irreducible $\mathbf{F}G$-module. Note that

$$d_k^*: \mathrm{Hom}_{\mathbf{F}G}(P_{k-1}, M) \to \mathrm{Hom}_{\mathbf{F}G}(P_k, M)$$

is the zero-map. Indeed, if $\phi \in \mathrm{Hom}_{FG}(P_{k-1}, M)$ then

$$d_k^*(\phi)(x) = \phi \circ d_k(x) \subseteq \phi(\ker d_{k-1}) \subseteq \phi(JP_{k-1}) \subseteq JM = 0$$

for all $x \in P_k$. This shows that

$$H^k(G, M) = \mathrm{Hom}_{FG}(P_k, M)$$

for all $k \in \mathbb{N}$. Define $E = \mathrm{Hom}_{FG}(M, M)$. Then we have

$$H^k(G, M) = \mathrm{Hom}_{FG}(P_k, M) \cong E^{e(P_k, M)}$$

and hence

$$\dim_F(H^k(G, M)) = \dim_F E \cdot e(P_k, M).$$

Note further that

$$\mathrm{Hom}_{FG}(P_k, M) \cong \mathrm{Hom}_{FG}(\ker d_{k-1}, M).$$

This is because every $\phi \in \mathrm{Hom}_{FG}(P_k, M)$ factors through JP_k, and since $\ker d_k \subseteq JP_k$, ϕ factors through $\ker d_k$. Since $P_k/\ker d_k \cong d_k(P_k) = \ker d_{k-1}$, the claim follows. Thus we also have

$$\dim_F(H^k(G, M)) = \dim_F E \cdot e(\ker d_{k-1}, M).$$

Let $A = \ker d_1$. We compare the exact sequences

$$0 \to A \to P_1 \xrightarrow{d_1} P_0 \xrightarrow{\epsilon} \mathbf{F} \to 0$$

and

$$0 \to \bar{N} \to \mathbf{F}G^n \xrightarrow{\partial_1} \mathbf{F}G \to \mathbf{F} \to 0,$$

where $n = d(F)$. By Schanuel's lemma

$$\bar{N} \oplus P_1 \cong A \oplus \mathbf{F}G^{n-1} \oplus P_0.$$

If we apply $\mathrm{Hom}_{FG}(-, M)$ we obtain

$$E^{e(\bar{N}, M)} \oplus E^{e(P_1, M)} \cong E^{e(A, M)} \oplus E^{(n-1) \cdot e(\mathbf{F}G, M)} \oplus E^{e(P_0, M)}.$$

Counting dimensions we obtain

$$e(\bar{N}, M) + e(P_1, M) = e(A, M) + (n-1) \cdot e(\mathbf{F}G, M) + e(P_0, M).$$

Thus

$$\frac{e(\bar{N}, M)}{e(\mathbf{F}G, M)}$$

is equal to

$$\frac{\dim_F H^2(G, M)}{\dim_F E \cdot e(FG, M)} - \frac{\dim_F H^1(G, M)}{\dim_F E \cdot e(FG, M)} + (n-1) + \frac{\dim_F H^0(G, M)}{\dim_F E \cdot e(FG, M)}.$$

Now note that $\dim_F E \cdot e(FG, M) = \dim_F M$, because $\text{Hom}_{FG}(FG, M) \cong M$. Thus we see that

$$\frac{e(\bar{N}, M)}{e(FG, M)}$$

is equal to

$$\frac{\dim_F H^2(G, M)}{\dim_F M} - \frac{\dim_F H^1(G, M)}{\dim_F M} + (n-1) + \frac{\dim_F H^0(G, M)}{\dim_F M}.$$

Lemma 6.16 now implies that the formula in the statement of Theorem 6.15 does hold. □

Denote by $\pi(G)$ the set of primes that divide the order of the group G. Let F/N be a finite presentation for G. A prime $p \in \pi(G)$ is called a *relation prime for F/N* if

$$d_{\mathbb{Z}G}(N/[N,N]) = d_{\mathbb{Z}_p G}(\mathbb{Z}_p \otimes (N/[N,N])) = d_{\mathbb{Z}_p G}(N/N^p[N,N]).$$

Here $\mathbb{Z}_p$ denotes the field with p elements. The relation prime notion is presentation independent: if p is a relation prime for some finite presentation then it is a relation prime for all finite presentations. Thus we can speak of the set of relation primes for the group G. It is known that this set of relation primes is non-empty for a non-trivial group. Proofs for the statements made on relation primes can be found in [Gru76]. As a consequence we obtain the following result which gives a tool for computing the rank of the relation module.

Corollary 6.17 $d_G(N/[N,N]) = \max\{d_G(N/N^p[N,N]) \mid p \in \pi(G)\}$, and $d_G(N/N^p[N,N])$ is computed as in Theorem 6.15.

Corollary 6.18 *Let F/N be a finite presentation of a finite group G. Then the difference*

$$d_G(N/[N,N]) - d(F)$$

is a group invariant.

Proof: The difference in the statement depends only on G by Theorem 6.15. □

This result does not extend to infinite groups. The trefoil knot group presented by

$$\langle x, y \mid x^2 = y^3 \rangle$$

has 2-generator presentations whose relation module is not generated by a single element. See Dunwoody [Dun72].

Relation gaps over group algebras other than the integral group ring can easily be constructed. Note that the group algebra $\mathbb{Q}G$ is semi-simple by Maschke's theorem. Since $\mathbb{Q}$ is a submodule of $\mathbb{Q}G$, it is a direct summand. Hence $\mathbb{Q}$ is projective and $0 \to \mathbb{Q} \to \mathbb{Q} \to 0$ is a projective resolution. It follows that $H^i(G, M) = 0, i \geq 1$, for all $\mathbb{Q}G$-modules M. Thus $d_{\mathbb{Q}G}(\mathbb{Q} \otimes (N/[N,N])) = d(F)$. Hence, if F/N is not balanced, then this presentation has a relation gap over $\mathbb{Q}$. For example, if $G = \mathbb{Z}_2 \times \mathbb{Z}_2$, then every presentation of G has a relation gap over $\mathbb{Q}$.

We give another consequence of Theorem 6.15 due to Wamsley [Wam70]. It implies that a finite nilpotent group that is not efficient has a presentation with a relation gap. Recall that non-efficient finite solvable groups are known to exist (Swan [I], [Sw65]).

Corollary 6.19 *(Wamsley) Let F/N be a finite presentation of a finite nilpotent group G. Then F/N is proficient.*

The corollary follows from the fact that in case of a finite nilpotent group G we have $H^i(G, M) = 0, i \geq 1$, for every irreducible $\mathbb{Z}_p G$-module $M \neq \mathbb{Z}_p$, $p \in \pi(G)$. If p is a relation prime for F/N, then it follows from Theorem 6.15 that

$$d_G(N/[N,N]) = \dim_{\mathbb{Z}_p} H^2(G, \mathbb{Z}_p) - \dim_{\mathbb{Z}_p} H^1(G, \mathbb{Z}_p) + d(F)$$
$$= d(F) - \dim_{\mathbb{Z}_p} H_1(G, \mathbb{Z}_p) + \dim_{\mathbb{Z}_p} H_2(G, \mathbb{Z}_p).$$

Hence F/N is proficient by Lemma 6.2.

A group G is called *balanced* if it has a finite presentation F/N where $d(F) = d_F(N)$. In his book [Joh97], Chapter 7, D. L. Johnson calls finite balanced groups *interesting groups*. We quote from his book:"While our current (1990) stock of interesting groups with 2-generators is very large, the list of those needing 3 generators is fairly short." At present we do not know any interesting groups requiring more than 3 generators.

Theorem 6.20 *Suppose there exists $n \geq 4$ so that there does not exist an interesting group (finite and balanced) which requires at least n generators. Then there exists a presentation F/N of a finite group G that has a relation gap.*

This result follows from a result of Gruenberg [Gru80]. He showed that if G is a finite proficient group, and $H_1(G) = H_2(G) = 0$, then G^k is also proficient for any $k \geq 1$. Let G be the binary icosahedral group. The first and second homology of G are trivial and G has a proficient 2-generator 2-relator presentation. Choose k large enough so that $d(G^k) \geq n$. Then there exists a presentation F/N of G^k, where F is free of rank n. Since G^k has a proficient presentation and $d(F) - d_G(N/[N,N])$ is a group invariant by Corollary 6.18, it follows that $d_G(N/[N,N]) = n$. But $d_F(N) > n$, since otherwise F/N would be balanced. Hence F/N has a relation gap. □

7

On the Relation Gap Problem for Free Products

Cynthia Hog-Angeloni and Wolfgang Metzler

7.1 Introduction

Finite relation gaps have been conjectured in several cases where they don't exist after all, see [I], page 50. A notorious open example are the groups $G * H, G = \mathbb{Z}_k \times \mathbb{Z}, H = \mathbb{Z}_\ell \times \mathbb{Z}$ with $(k, \ell) = 1, k, \ell > 1$, which were mentioned by D. B. A. Epstein in [Eps61]. In this chapter we give a partial answer for cases of semisplit presentations: *We prove that for "natural" generators a, b of G, α, β of H, except for the case of one split and two unsplit relators which we have to leave open, any semisplit presentation of the Epstein groups needs at least 4 defining relations (section 3)*. The terminology and facts about semisplit presentations of free products are taken up from [I], Chapter XII. But for the integral relation module K. W. Gruenberg and P. Linnell [GrLi08] have shown that already three generators suffice, one such generating system we give in Section 2. In order to establish a finite *relation gap* for the Epstein groups we would have to show, that such a difference can also be established for all semisplit presentations in the presence of additional generators which are trivial in $G * H$. This is a topic for further research.

One main technique is an inductive process to obtain split consequences of semisplit presentations, the Flower Theorem in Section 2. We also mention (partial) results on other free products and that our results in Section 3 have some similarity with Proposition 3 in M. R. Bridson and M. Tweedale [BrTw07].

In Section 5 we show moreover that, surprisingly, in passing to coefficients in the group ring over some *field* **F**, any set of generators of the relation module of a semisplit presentation is elementarily equivalent to a set of generators which do live over the factors. If $\mathbf{F} = \mathbb{Q}$, one may speak of "*rational relation gaps*" which occur for instance in the cases $\mathbb{Z}_k \times \mathbb{Z}_k$ and $\mathbb{Z}_k \times \mathbb{Z}, k > 1$.

In Section 6, we finally give a condensed version of the proof of Theorem

3.4 in [I], Chapter XII, on semisplit Q^{**}-transformations, as its idea is related to the one of the Flower Theorem.

As commutators of relations become trivial when passing from presentations to relation modules, and as two presentations, whose defining relations just differ by commutators of relations, mark the potential distinction between sh-type and Q^{**}-equivalence of 2-complexes (see [I], page 370 for details), there is good reason to expect that further progress on one of these problems will also be helpful for the other one.

In analogy to Chapter 2 we thank our Master students Janina Körner (Frankfurt/Main) and Christopher Muth (Mainz). Their theses and the cooperation with each of them inspired this chapter.

7.2 Semisplit presentations, flower relations and integral flower chains

In [I], page 373, we proved the following:

Theorem 7.1 *In every presentation class of a finitely presentable free product $G * H$ there exists a finite presentation of the form*
$\mathcal{P} = \langle a_*, b_* \mid R_*, S_*, u_* v_*^{-1} \rangle$ *where the R_* and the u_* are words in the free group $F(a_*)$, and the S_* and the v_* are words in the free group $F(b_*)$; G is the group defined by $\langle a_* \mid R_*, u_* \rangle$ and H is defined by $\langle b_* \mid S_*, v_* \rangle$.*

Here and throughout this chapter the index $*$ shall denote an index i which runs through a set $\{1, \ldots, n\}$ for some n. A presentation $\mathcal{P}$ as in the above theorem was called a *semisplit* presentation, the R_*, S_* *split* relators of $\mathcal{P}$, $T_* := u_* v_*^{-1}$ the *unsplit* ones. By U_0 we denote the subgroup $\text{gp}_{F(a_*)}(u_*)$ of $F(a_*)$ generated by the u_*; V_0 denotes $\text{gp}_{F(b_*)}(v_*)$.

Let G_1 be the group presented by $\langle a_* \mid R_* \rangle$ and H_1 the group presented by $\langle b_* \mid S_* \rangle$; also $U_1 := \text{gp}_{G_1}(u_*)$, $V_1 := \text{gp}_{H_1}(v_*)$. Overlining a word in $F(a_*) * F(b_*)$ denotes its projection to $G_1 * H_1$. Throughout this chapter we assume that u_*, v_* is a basis for U_0, V_0 respectively; otherwise using Nielsen moves we could reduce the number of unsplit relators T_*.

The notion of a *semisplit presentation* is best illustrated by the example from [I], p. 374, which will serve as a guide through this section, i.e. we will work through this example step by step.

(Guiding Example) A presentation of $(\mathbb{Z}_2 \times \mathbb{Z}_2) * (\mathbb{Z}_3 \times \mathbb{Z}_3)$ with few relations:

Let $\mathcal{P} = \langle a, b, \alpha, \beta \mid a^2, aba^{-1} = b^3, \alpha^3, \alpha\beta\alpha^{-1} = \beta^4, u = v \rangle$, where u abbreviates b^2 and $v\beta^3$. A little computation shows that G_1 is $\mathbb{Z}_2 \ltimes \mathbb{Z}_8$, H_1 is $\mathbb{Z}_3 \ltimes \mathbb{Z}_{63}$. One sees that u has order 4 in G_1 and v has order 21 in H_1. As these numbers are coprime, the relation $u = v$ implies that u and v are both trivial. We conclude that $\mathcal{P}$ is a semisplit presentation of the free product $(\mathbb{Z}_2 \times \mathbb{Z}_2) * (\mathbb{Z}_3 \times \mathbb{Z}_3)$. □

In (small) examples as here the generators are called a and b for G, α and β for H, whereas for the general setting we use a_* for those of G and b_* for those of H.

Consider the natural bijection $\varphi : \{u_*\} \to \{v_*\}$ sending u_i to "its twin" v_i. φ induces an isomorphism φ_1 from U_1 to V_1 if and only if for every word in the u_i which is 1 in U_1, the corresponding word in the v_i is 1 in V_1 and vice versa. But if φ_1 were an isomorphism, then $\mathcal{P}$ would present the free product $G_1 * H_1$ with amalgamated subgroups U_1 and V_1. In particular, G_1 and H_1 would embed into G contradicting the assumption $1 \neq \overline{u}_* \in (ker(G_1 \to G))$.

We have proved the following:

Lemma 7.2 *Up to interchanging the roles of G and H and with the above assumptions, there exists a word $W(u_*)$ such that $\overline{W(u_*)} = 1$, but the twin word $W(v_*)$ which replaces every u_* by v_* projects nontrivially to H_1.*

$W(v_*)$ as in Lemma 7.2 is called a *flower-relation* of $\mathcal{P}$: Let $K_\mathcal{P}^2$ be the standard 2-complex of $\mathcal{P}$. Think of a small disc with boundary labelled $W(u_*) = u_{i_1}^{\varepsilon_1} \ldots u_{i_n}^{\varepsilon_n}$ and inside the disc a contraction via the 2-cells corresponding to R_*. At each $u_{i_j}^{\varepsilon_j}$ we attach a "petal" T_{i_j}. We thus obtain a diagram with the shape of a flower picturing a map $f : D^2 \to K_\mathcal{P}^2$, such that $f|S^1$ reads the flower relation $W(v_*)$.

Here are examples of flower relations of the presentation investigated in our **Guiding Example**:

From the above observation that u^4 is a consequence of R_1 and R_2, and that v^{21} is a consequence of S_1 and S_2, we obtain the flower relations v^4 and u^{21}.

Notice that $\langle a, b \mid R_*, u^{21} \rangle$ (resp. $\langle \alpha, \beta \mid S_*, v^4 \rangle$) are presentations for G and H, just as $\langle a, b \mid R_*, u \rangle$ (resp. $\langle \alpha, \beta \mid S_*, v \rangle$) are.

In general there will be possible iterations of this process: If W_1 is a flower relation of $\mathcal{P}$, $\mathcal{P} \cup W_1$ may admit another flower relation etc.

That after finitely many steps every split consequence of $\mathcal{P}$ can be obtained this way, is the content of the following *Flower Theorem*:

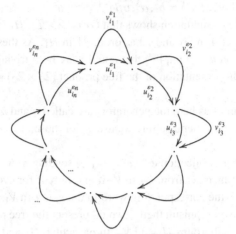

Figure 7.1 Flower Relation

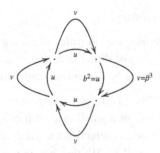

Figure 7.2 Flower relation v^4

Theorem 7.3 *Let a free product $G * H$ with a semisplit presentation $\mathcal{P} = \langle a_*, b_* \mid R_*(a_*), S_*(b_*), u_* v_*^{-1} \rangle$ be given and let W be a split consequence of $\mathcal{P}$. Then there exists a split presentation*

$$Q = \langle a_*, b_* \mid R_*(a_*), S_*(b_*), W_1, \ldots, W_m \rangle$$

where W_{j+1} is a flower relation over $\mathcal{P} \cup \{W_1, \ldots, W_j\}$ and $W_m = W$.

Proof: The idea is simply to express each u_*, v_* as a consequence of the relations of $G * H$ and at the intermediate level $G_1 * H_1$ transform the conjugating elements into some normal form as if $G * H$ were a free product $G_1 * H_1$ with amalgamated subgroups U_1 and V_1. Then the desired flower relations arise all by themselves.

On the Relation Gap Problem for Free Products 153

Inductively, G_k is defined to be G_{k-1}, if W_{k-1} is red, and to be $G_{k-1} \cup W_{k-1}$, if W_{k-1} is green, analogously for H_k.

As u_*, v_* project to $1 \in G * H$, in $F(a_*) * F(b_*)$ there are products P_{u_*}, P_{v_*} of conjugates of the unsplit relators $T_* = u_* v_*^{-1}$ such that, say,

$$\prod_{i=1}^{m} w_i T_{j_i}^{\varepsilon_i} w_i^{-1} =: P_{u_*} \text{ satisfies } \overline{P}_{u_*} = \overline{u}_* \in G_1 * H_1 \qquad (7.1)$$

where $\varepsilon_i \in \{\pm 1\}$, $w_i \in F(a_*) * F(b_*)$.

W.l.o.g. we assume that no subproduct $(w_i T_{j_i}^{\varepsilon_i} w_i^{-1})(w_{i+1} T_{j_{i+1}}^{\varepsilon_i} w_{i+1}^{-1}) \cdots$ is trivial in $F(a_*) * F(b_*)$; in the language of diagrams this means that no subdiagram can be squeezed off.

The length $|\cdot|$ of a word $w \in F(a_*) * F(b_*)$ is always taken in $F(a_*) * F(b_*)$, and that of its projection $\overline{w} \in G_1 * H_1$, in $G_1 * H_1$.

Let $\ell = \sum_{u_*} |P_{u_*}| + \sum_{v_*} |P_{v_*}|$ be the sum of the lengths of all products P_*.

In general, for $P \in \{P_{u_*}, P_{v_*}\}$ the length $|P|$ may be big – P consists of alternating green segments in $F(a_*)$ and red segments in $F(b_*)$ – but $\overline{P}$ has length 1 in $G_1 * H_1$. (Think of the a_* as being green and the b_* as being red).

By the Normal Form Theorem, P then contains some, say, green segment that projects to $1 \in G_1$. The idea now is to show

i) P in fact contains some such segment which lies entirely in U_0 or V_0.
ii) Inserting an appropriate flower relation after such a segment can be achieved preserving condition (7.1) and trades a, say, green segment (which is a consequence of the R_* only) for the twin red segment, thus lowering ℓ.

We start with ii):

If $|P| \leq 1$, we have $P = 1$ - this is because $\langle a_*, b_* \mid T_* \rangle$ *is* a free product with amalgamation into which $F(a_*)$ and $F(b_*)$ embed - and thus that the corresponding $\overline{u}_* = 1$.

When $\ell = 0$, then $|P| = 0$ for all $P \in \{P_{u_*}, P_{v_*}\}$, in other words $1 = P = \overline{P} = \overline{u}_*$ (resp. $\overline{v}_*$) $\in G_k * H_k$. Hence $G_k * H_k = G * H$ and we have a presentation with flower relations W_* which dispenses with the unsplit relations T_*.

Thus we assume some $|P| > |\overline{P}|$.

If P contains a segment $W(u_*) = u_{i_1}^{\varepsilon_1} \ldots u_{i_n}^{\varepsilon_n} \in U_0$ in one of the $P \in \{P_{u_*}, P_{v_*}\}$ which projects to 1, we substitute it by its twin $W(v_*) \in V_0$ by inserting after $W(u_*)$ the following product of conjugates of $T_{i_j} = u_{i_j}^{-\varepsilon_j} v_{i_j}^{\varepsilon_j}$

$$T_{i_n} \cdot \left(v_{i_n}^{-\varepsilon_n} T_{i_{n-1}} v_{i_n}^{\varepsilon_n}\right) \cdots \left(v_{i_n}^{-\varepsilon_n} \cdots v_{i_1}^{-\varepsilon_1} T_{i_1} v_{i_1}^{\varepsilon_1} \cdots v_{i_n}^{\varepsilon_n}\right) = W(u_*)^{-1} W(v_*) \quad (7.2)$$

thus shortening the length $|P|$. Either $\overline{v_{i_1}^{\varepsilon_1} \ldots v_{i_n}^{\varepsilon_n}} = 1$ as well, in which case $\overline{P}$ isn't changed; or $\overline{v_{i_1}^{\varepsilon_1} \ldots v_{i_n}^{\varepsilon_n}} \neq 1$ in which case we introduce the flower relation $W(v_*)$ defining H_2 (we set $G_2 := G_1$). Equation (7.1) now holds over $G_2 * H_2$: The image of P in $G_2 * H_2$ is now the projection of u_* to G_2 (which we continue to denote by $\overline{u}_*$) and by shortening the length of P, ℓ has decreased.

If we find another segment in U_0 or V_0 we proceed inductively. After the k–th step we have

(1) $$P_{u_*} = \prod_{i=1}^{m} w_i T_{j_i}^{\varepsilon_i} w_i^{-1} \text{ and } \overline{P}_{u_*} = \overline{u}_* \in G_k * H_k$$

where no subproduct $(w_i T_{j_i}^{\varepsilon_i} w_i^{-1})(w_{i+1} T_{j_{i+1}}^{\varepsilon_i} w_{i+1}^{-1}) \cdots$ is trivial in $F(a_*) * F(b_*)$, $\varepsilon_* \in \{\pm 1\}$, $w_i \in F(a_*) * F(b_*)$.

As a warm-up case for step i), assume that all the conjugating elements of P (and thus P itself) were already in $U_0 * V_0$. By the Normal Form Theorem, P contains a, say, green segment which projects to 1. By the assumption of the case at hand, this segment is of the form $W(u_*) = u_{i_1}^{\varepsilon_1} \ldots u_{i_n}^{\varepsilon_n} \in U_0$, thus we can proceed as described above.

For the general case we first perform a normalization procedure to the conjugating elements w_i by projecting their segments to G_1 (resp. H_1) and there express each of them as a product of a coset representative $[G_1 : U_1]$ (resp. $[H_1 : V_1]$) with an element of $U_1 \leq G_1$ (resp. $V_1 \leq H_1$). These factors are then lifted back to the free group; the coset representative to some $\widehat{g}_* \in F(a_*)$ (or $\widehat{h}_* \in F(b_*)$), the element of $U_1 \leq G_1$ (or $V_1 \leq H_1$) to some word $x_i(u_*)$ (or $y_i(v_*)$) in the u_* (or v_*). We furthermore require $\widehat{g}_{i,j} = 1$ if $\overline{\widehat{g}}_{i,j}$ represents U_1 and that $1 \in U_1$ is lifted to $1 \in U_0$. In doing so, (7.1) is unchanged and ℓ can only decrease.

Definition: A word $w_i = g_{i_1} h_{i_1} \ldots g_{i_n} h_{i_n}$ in $G_k * H_k$ is called *reduced* if it satisfies the following condition:
w_i equals $\widehat{w}_i x_i(u_*)$ or $\widehat{w}_i y_i(v_*)$ where $\widehat{w}_i$ is of the form $\widehat{g}_{i_1} \widehat{h}_{i_1} \ldots \widehat{g}_{i_n} \widehat{h}_{i_n}$ – $\widehat{w}_i$ could start or end in either $F(a_*)$ or $F(b_*)$ – with $\widehat{g}_* \in F(a_*)$ such that $\overline{\widehat{g}}_*$ is a nontrivial element (i.e. $\overline{\widehat{g}}_* \notin U_1$) of a system of coset representatives $[G_1 : U_1]$; similarly $\widehat{h}_* \in F(b_*)$, and the $x_i(u_*)$ resp. $y_i(v_*)$ are words in the u_* resp. v_*. Note the colour change after $\widehat{w}_i$.

We now set out to normalize the conjugating elements w_i in (1). Working from left to right we assume that the first $2j-2$ segments of $w = w_i$ are already as desired, i.e. $w = \widehat{g_1}\widehat{h_1}\cdots\widehat{h_{j-1}}g_jh_j\cdots g_sh_s$ with no segment trivial. Then
$$w = \widehat{g_1}\widehat{h_1}\cdots\widehat{g_j}u_{i_1}^{\varepsilon_1}\cdots u_{i_t}^{\varepsilon_t}h_j\cdots g_sh_s$$
$$= \widehat{g_1}\widehat{h_1}\cdots\widehat{g_j}u_{i_1}^{\varepsilon_1}\cdots u_{i_t}^{\varepsilon_t}(v_{i_1}^{\varepsilon_1}\cdots v_{i_t}^{\varepsilon_t})^{-1}(v_{i_1}^{\varepsilon_1}\cdots v_{i_t}^{\varepsilon_t})h_j\cdots g_sh_s \text{ with } \varepsilon_* = \pm 1.$$

We set $h'_j := (v_{i_1}^{\varepsilon_1}\cdots v_{i_t}^{\varepsilon_t})h_j$, $w' = \widehat{g_1}\widehat{h_1}\cdots\widehat{g_j}$ and $w'' := g_{j+1}\cdots g_sh_s$ and further rewrite w as $w = (w'T_{i_t}w'^{-1})\cdot$
$$\left(w'v_{i_1}^{\varepsilon_1}T_{i_2}v_{i_1}^{-\varepsilon_1}w'^{-1}\right)\cdots\left(w'v_{i_1}^{\varepsilon_1}\cdots v_{i_{t-1}}^{\varepsilon_{t-1}}T_{i_t}v_{i_{t-1}}^{-\varepsilon_{t-1}}\cdots v_{i_1}^{-\varepsilon_1}w'^{-1}\right)\right)w'h'_jw''$$
with $T_{i_j} = u_{i_j}^{\varepsilon_j}v_{i_j}^{-\varepsilon_j}$.

P is unchanged, but the new conjugating element $w'h'_jw''$ has the first $2j-1$ segments in the desired form, and the conjugating elements of the additional T_{i_j} are reduced. If $\widehat{g_j} = 1$, the next step is done with $\widehat{h_{j-1}}h'_j$ rather than h'_j.

This normalization of the conjugating elements achieves the following:

(2) Segments $g_{i,*}$ in the conjugating elements with $\overline{g}_{i,*} = 1$ have been eliminated.

(3) At the "suture" $\widehat{w}_i^{-1}\cdot\widehat{w}_{i+1}$ cancellation can take place only freely, $|\widehat{w}_i^{-1}\cdot\widehat{w}_{i+1}|_{F(a_*)*F(b_*)} = |\widehat{w}_i^{-1}\cdot w_{i+1}|_{G_1*H_1}$ and every segment of $\widehat{w}_i^{-1}\cdot\widehat{w}_{i+1}$ projects to a coset $\neq U_1$.

Define $\sigma_0 := \widehat{w}_1$, $\sigma_m := \widehat{w}_m^{-1}$, $\sigma_i = \widehat{w}_i^{-1}\cdot\widehat{w}_{i+1}$ for $1 \leq i \leq m-1$ and for $1 \leq i \leq m$ let $\tau_i \in U_0 * V_0$ be what is left in between, i.e. $x_i(u_*)T_{j_i}^{\varepsilon_i}x_i^{-1}(u_*)$ resp. $y_i(v_*)T_{j_i}^{\varepsilon_i}y_i^{-1}(u_*)$.

Then $P = \sigma_0\tau_1\sigma_1\tau_2\cdots\tau_m\sigma_m$ where $\tau_* \in U_0 * V_0$ fulfills $2 \leq |\tau_*| \leq 3$. Omit the σ_* that are empty, and lump the ensuing maximal chains $\tau_j\tau_{j+1}\cdots$ together into subproducts Q_*. By (7.1), $Q_* \neq 1$. As $\langle u_*, v_* \mid T_*\rangle \approx F(u_*)$, $|Q_*| > 1$.

The segment, say $\widehat{g}_{i,*}x_{i,*}(u_*)x'_*(u_*)$ at the transition from σ_i to τ_i, resp. from τ_i to σ_{i+1}, doesn't project to 1 either.

Thus there must be some segment *within* some Q_* that projects to 1. As $Q_* \in U_0 * V_0$, we are done. □

Note that Theorem 7.3 can be generalized to a *partial semisplit presentation*, whose relators can be completed to a semisplit presentation of $G * H$ but may not be enough to present $G * H$ themselves.

In [Kör15], J. KÖRNER has proved such a generalization in a more geometric version of the proof for Theorem 7.3.

We also want to draw attention to the fact that after reductions the diagram may show local complications which are not seen in Figure 7.1 (see Figure 7.3).

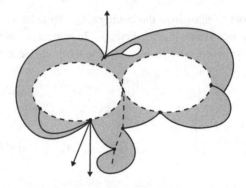

Figure 7.3 Flower with singularities

The petals are shaded and *separation points* are marked where the colour of their boundary segments changes. Each two separation points can be connected by a path of each colour which is trivial in the corresponding factor of $G * H$. Split relations may result from previous stages and may grow together in later ones.

We now pass from presentations to chain complexes: For a given presentation we model the standard 2-complex $K_{\mathcal{P}}^2$ and look at the chain complex of its universal covering $C_2(\widetilde{K}) \xrightarrow{\partial_2} C_1(\widetilde{K}) \xrightarrow{\partial_1} C_0(\widetilde{K})$ (which is exact at $C_1(\widetilde{K})$). As $\pi_1(\widetilde{K}^1) = N$, the relation module $\mathbf{N}_{\mathcal{P}}$ of $\mathcal{P}$ is $H_1(\widetilde{K}^1)$ and is thus generated by the rows of the Reidemeister-Fox-matrix A describing the second boundary homomorphism ∂_2 with respect to canonical bases: The row vectors of A are the *Reidemeister-Fox-derivative*s of the relators taken over $\mathbb{Z}(G * H)$.

Guiding Example The Reidemeister-Fox-matrix A for $\mathcal{P}$ has four *split* rows $\widetilde{R}_*, \widetilde{S}_*$ corresponding to the split relators R_*, S_* and one unsplit row $\widetilde{T}_*$; altogether these generate $\mathbf{N}_{\mathcal{P}}$:

	$\widetilde{a}$	$\widetilde{b}$	$\widetilde{\alpha}$	$\widetilde{\beta}$
$\widetilde{R_1}$	$1+a$	0	0	0
$\widetilde{R_2}$	$1-b$	$a-(b+2)$	0	0
$\widetilde{S_1}$	0	0	$1+\alpha+\alpha^2$	0
$\widetilde{S_2}$	0	0	$1-\beta$	$\alpha-(2+\beta+\beta^2)$
$\widetilde{T}$	0	$1+b$	0	$-(1+\beta+\beta^2)$

Here, the vector $\widetilde{T}$ is of the form $\left(\widetilde{u} \mid \widetilde{v}\right)$ where $\widetilde{u}$ is identified with $\left(\widetilde{u} \mid 0\right)$ and $\widetilde{v}$ with $\left(0 \mid \widetilde{v}\right)$. □

Flower relations give rise to "flower chains" which are $\mathbb{Z}$-linear combinations of chains corresponding to the "halves" u_i resp. v_i of the unsplit relators T_i and – as opposed to the T_* – live over the factors.

Definition: A *flower chain* is an element of $C_1(\widetilde{K})$ of the form $\left(0 \ldots 0 \mid \sum_i q_i \widetilde{v}_i\right)$ where the q_i are elements in $\mathbb{Z}$ and the *twin chain* $\left(\sum_i q_i \widetilde{u}_i \mid 0 \ldots 0\right)$ is a $\mathbb{Z}G$-linear combination of the $\widetilde{R}_*$; or the same with the roles of G and H interchanged.

Proposition: The Reidemeister-Fox-derivative over $\mathbb{Z}(G * H)$ of a flower relation is a flower chain.

Proof: Recall that the u_*, v_* project to $1 \in G * H$, hence the boundary chain of a flower relation, say, $W(v_*) = v_{i_1}^{\varepsilon_1} \ldots v_{i_n}^{\varepsilon_n}$ with $\varepsilon_* \in \{\pm 1\}$ is simply $\varepsilon_1 (0, \ldots, 0|\widetilde{v}_{i_1}) + \ldots + \varepsilon_n (0, \ldots, 0|\widetilde{v}_{i_n})$ and thus of the desired form $\left(0 \ldots 0 \mid \sum_i q_i \widetilde{v}_i\right)$. As $\overline{W}(u_*) = 1$, the twin chain $\left(\sum_i q_i \widetilde{u}_i \mid 0 \ldots 0\right)$ is the boundary of a $\mathbb{Z}G$-linear. combination of $\widetilde{R}_*$. □

Guiding Example The flower chain associated to the flower relation of Figure 7.2:
The flower chain associated to the flower relation v^4 is $\left(\begin{array}{cc|c} 0 & 0 & 4\widetilde{v}\end{array}\right) = \left(\begin{array}{cc|cc} 0 & 0 & 0 & 4(1+\beta+\beta^2)\end{array}\right)$. The twin chain $\left(\begin{array}{c|cc} 4\widetilde{u} & 0 & 0\end{array}\right) = \left(\begin{array}{cc|cc} 0 & 4(1+b) & 0 & 0\end{array}\right)$ is a linear combination of $\left(\begin{array}{c|cc}\widetilde{R}_1 & 0 & 0\end{array}\right)$ and $\left(\begin{array}{c|cc}\widetilde{R}_2 & 0 & 0\end{array}\right)$ with coefficients in $\mathbb{Z}G$. (Express u^4 as a product of conjugates of R_1, R_2 and take the derivatives). □

Note that in general nonsplit relators may have split derivatives.

As a second example we treat the *Epstein groups* with presentation $\mathcal{P} = \langle a, b, \alpha, \beta \mid a^k, [a,b], \alpha^\ell, [\alpha, \beta] \rangle$ for $(k, \ell) = 1$. Its relation module is generated by

158   Cynthia Hog-Angeloni and Wolfgang Metzler

$$
\begin{array}{rcccccc}
 & & \widetilde{a} & \widetilde{b} & \widetilde{\alpha} & \widetilde{\beta} & \\
\widetilde{u}_1 & = & (\Sigma_a & 0 & | & 0 & 0 &) \\
\widetilde{u}_2 & = & (1-b & a-1 & | & 0 & 0 &) \\
\widetilde{v}_1 & = & (0 & 0 & | & \Sigma_\alpha & 0 &) \\
\widetilde{v}_2 & = & (0 & 0 & | & 1-\beta & \alpha-1 &)
\end{array}
$$

where $\Sigma_a = \sum_{i=0}^{k-1} a^i, \Sigma_\alpha = \sum_{i=0}^{\ell-1} \alpha^i$.

But we can generate the relation module already by three elements:
Define $\widetilde{R} := k\widetilde{u}_2 + b\widetilde{u}_1$ and $\widetilde{S} := \ell\widetilde{v}_2 + \beta\widetilde{v}_1$. Moreover define $\widetilde{T} := \widetilde{u}_2 - \widetilde{v}_2$.
A direct computation shows
$\Sigma_a \widetilde{R} = b\Sigma_a\widetilde{u}_1 + k\Sigma_a\widetilde{u}_2 = (\ k\Sigma_a(b+(1-b)) \ | \ 0 \ ) = k\widetilde{u}_1$.
$(k - b\Sigma_a)\widetilde{R} = k\widetilde{R} - bk\widetilde{u}_1 = k^2\widetilde{u}_2 + bk\widetilde{u}_1 - bk\widetilde{u}_1 = k^2\widetilde{u}_2$.
analogously $(\ell - \beta\Sigma_\alpha)\widetilde{S} = \ell^2\widetilde{v}_2$.
Hence $\widetilde{R}, \widetilde{S}$ and $\widetilde{T}$ generate $k^2\widetilde{u}_2$ and $\ell^2\widetilde{u}_2$ and thus $\widetilde{u}_2$, as k and ℓ are coprime. Now it is immediate that $\widetilde{R}, \widetilde{S}$ and $\widetilde{T}$ are three generators of the integral relation module of $\mathcal{P}$.

In [GrLi08], KARL W. GRUENBERG and PETER A. LINNELL give formulae in various situations for the minimum number of elements required to generate a $\mathbb{Z}G$-module of the form $\oplus_{i=1}^n M_i \otimes_{\mathbb{Z}G_i} \mathbb{Z}G$, G a free product $G_1 * \ldots * G_n$.

$\widetilde{R}, \widetilde{S}$ and $\widetilde{T}$ look as if they might be lifted to a semisplit presentation with two split relators and a nonsplit one. Explicit generators like in the case of the Epstein groups but for more than two coprime factors k, ℓ have recently been given by Wajid Mannan [Man16].

7.3 Applications to the Epstein Groups

We now apply the Flower Theorem to show that in three of four cases a free product π with presentation $\langle a, b \mid a^k, [a,b] \rangle * \langle \alpha, \beta \mid \alpha^\ell, [\alpha,\beta] \rangle$ for $(k, \ell) = 1$ doesn't admit a semisplit presentation with 3 defining relators for the natural generators a, b, α, β. Less relators don't suffice because of the fact, that by [GrLi08] already the relation module needs that many module generators. Recall that by [I], [Re36] $G = (\mathbb{Z}_k \times \mathbb{Z})$ and $H = (\mathbb{Z}_\ell \times \mathbb{Z})$ aren't 1-relator groups and let us assume that we have a semisplit presentation $\mathcal{P}$ for π with 3 defining relators in a, b, α, β. We may further assume that there is no possibility of splitting the nonsplit ones by applying Nielsen operations. Last not least, we use that the R_i and u_i resp. S_i and v_i of all defining relators (split or nonsplit) present G resp. H. Then we have four cases:

Case i, $(i = 0, \ldots, 3)$: $\mathcal{P}$ has $3 - i$ split defining relators.
We show that the cases $0, 1, 3$ actually cannot occur:

Case 0: G or H would have to be a 1-relator group.

Case 1: Let the first and the second relation be split, $u = v$ be nonsplit. In this case, the split ones cannot be relations for the same factor G resp. H, as the other factor again would have to be a 1-relator group. Hence we assume that R refers to G and S to H. In order to get consequences on one side, the Flower-Theorem 7.3 will imply: An innermost flower with centre in G yields that either u is a consequence of R - and hence this case can be excluded - or u is a torsion element of order $\neq 1$ in the 1-relator group defined by R. Now by the theorem on torsion in 1-relator groups (see [KaMaSo60]) we have $R = U^m$, $u = wU^{\pm n}w^{-1}$, $(m, n) \in \mathbb{N}$, for some $U, w \in F(a, b)$. Hence G would also be a 1-relator group with defining relation $U^{(m,n)}$. The situation for an elementary flower in H is symmetric.

Case 3: For three nonsplit relators $u_1 = v_1$, $u_2 = v_2$, $u_3 = v_3$ which don't admit Nielsen reductions, the Flower-Theorem says that $\mathcal{P}$ doesn't yield any split consequences at all, as there is no start.

Case 2 we can't fully treat so far; subcases which substantiate our expectation, that case 2 also cannot occur, together with ideas how to proceed, are treated in the next section. There we also briefly consider the questions what may happen, if the restriction to the natural generators is dropped, and to others than the Epstein groups.

7.4 Ideas for further work

In case 2, i. e. two nonsplit and one split relator, one may start with a split relator $R \in F(a, b)$ and two Nielsen-independent nonsplit relators $u_2 = v_2$, $u_3 = v_3$; the nonsplit ones define a free product with amalgamation $F_1(a, b) *_A F_2(\alpha, \beta)$, A a free group of rank two generated by u_1, u_2 resp. v_1, v_2. Let N be the normal closure of R in F_1.

(4) If $A \cap N = \{1\}$, then a consequence of R in $F_1 *_A F_2$ is already a consequence of R in F_1.

A proof of this fact can be given by expressing such a consequence as a conjugate product of R in $F_1 *_A F_2$ and applying length reductions to Zieschang's normal forms for free products with amalgamation [Zie70], see also [I], [LySc77] from page 71 on. Applying (4) to case 2, one could remove u_1 and u_2 from the given presentation of $G * H$ by Q-*transformations*. We would be in the situation of a 1-relator group once more.

But (4) is only a subcase, in which A and N are "independent". Nevertheless many presentations of type 2 may be produced from a single one by applying semisplit Q-transformations (see [I]), page 374). We don't know so far, whether for one of the resulting presentations $A \cap N = \{1\}$ can be achieved, or, if – by other means – case 2 can be reduced to the remaining ones. As the amalgamating groups A vary, we are interested in length-notions which combine the behaviour by preparatory semisplit Q-moves and the one with free products with amalgamation.

In contrast to independence, we may look at the situation $N \subsetneq A$. In this case, R can be replaced by multiplication of an $u_1 (= v_1)$ and $u_2 (= v_2)$ - word to become its twin in F_2. Thus we can generate corresponding elements of G and H and in particular look for those of finite order k, ℓ or commutators. But so far we don't understand enough of the possibilities unlike in the relation module case, where even derivatives of elements with $(k, \ell) = 1$ can be realized.

With respect to elements of finite order we are also interested whether IRIS ANSHEL's thesis on 2-relator groups [Ans91] can be of use for case 2 like [KaMaSo60] for case 1. Of course, in case 2 we also consider ideas to lift module generators to normal generators of N.

Let us consider ideas how to establish a nontrivial finite relation gap for the Epstein Groups:

By [I], page 373, every presentation of $(\mathbb{Z}_k \times \mathbb{Z}) * (\mathbb{Z}_\ell \times \mathbb{Z})$ is Q**-equivalent to a semisplit one, and by further generator transformations we may assume that the free factors are generated by a, b resp. α, β and a finite number of generators for each factor, which become trivial when being projected to $\mathbb{Z}_k \times \mathbb{Z}$ resp. $\mathbb{Z}_\ell \times \mathbb{Z}$. It suffices to establish gaps for these situations.

Our computation at the end of section 7.2 shows that the relation module can be generated by less elements than expected, and any such presentation has at least one nonsplit relator. Now we have to argue in all cases where there is at least one split relator and a finite number of nonsplit ones. Certainly this needs specific properties of $(\mathbb{Z}_k \times \mathbb{Z}), (\mathbb{Z}_\ell \times \mathbb{Z})$. This time we may use all semisplit Q**-moves as preparatory means in order to reduce the number of cases.

So far we have concentrated on the Epstein-Groups. But there are other free product examples where a finite relation gap is conjectured to exist, see [I], [Ho-An88], with groups of type $(\mathbb{Z}_k \times \mathbb{Z}_{k\ell m}) * (\mathbb{Z}_\ell \times \mathbb{Z}_{k\ell m}) * (\mathbb{Z}_m \times \mathbb{Z}_{k\ell m})$.

These examples have the following history: Take three groups $\mathbb{Z}_k \times \mathbb{Z}_k$, $\mathbb{Z}_\ell \times \mathbb{Z}_\ell, \mathbb{Z}_m \times \mathbb{Z}_m$ with coprime indices and form their free product. The relation module of this group can be generated by 7 elements, see [I], page 50. W. METZLER has shown that these generators for the relation module can be lifted, i. e. there is in fact a presentation of the group in question with just 7 relations. In other cases, e. g. for $(\mathbb{Z}_k \times \mathbb{Z}_{k\ell m}) * (\mathbb{Z}_\ell \times \mathbb{Z}_{k\ell m}) * (\mathbb{Z}_m \times \mathbb{Z}_{k\ell m})$ see [I], [Ho-An88], the construction fails, so these are further candidates for a relation gap.

We close this section by **comparing our work with the one of M. BRIDSON and M. TWEEDALE**. In [BrTw07] they take up the study of the guiding example in section 7.2 of free product presentations for $(\mathbb{Z}_m \times \mathbb{Z}_m) * (\mathbb{Z}_n \times \mathbb{Z}_n)$ which "save" relators with respect to the sum of the values for the free factors. In these examples no relation gap shows up. Each free factor Q_m is an HNN-extension of $\mathbb{Z}_m \times \mathbb{Z}_m$ where a generator a for the first factor $\mathbb{Z}_m$ is conjugated with a stable letter t_m to become the generator of the second $\mathbb{Z}_m$ and likewise a generator α for the first factor $\mathbb{Z}_n$ is conjugated by t_n to become the generator of the second $\mathbb{Z}_n$. One thus gets presentations of a free product of HNN-extensions $\Gamma_{m,n} = Q_m * Q_n$ with generators a, α, t_m, t_n. In the case of m, n being two distinct primes, the calculations of [I], [HoLuMe85] yield presentations with four defining relations $\rho_m = t_m a t_m^{-1} \cdot a \cdot t_m a^{-1} t_m^{-1} \cdot a^{-m-1}, \rho_n = t_n \alpha t_n^{-1} \cdot \alpha \cdot t_n \alpha^{-1} t_n^{-1} \cdot \alpha^{-n-1}, a^m$ and α^n, such that the relation module is generated by the images of the three relations ρ_m, ρ_n and $a^m = \alpha^n$.

But now it is unknown whether the presentations have a relation gap or not. In a certain analogy to section 3, Bridson and Tweedale prove in [BrTw07] that there is no presentation of $\Gamma_{m,n}$ with generators a, α, t_m, t_n and defining relations ρ_m, ρ_n together with any third word r in a, α, t_m, t_n. Comparing with our cases, two of these relations are fixed; the third one is totally arbitrary. Instead of the torsion theorem for 1-relator groups, Bridson and Tweedale use J. Howie's theorem on torsion in free products of locally indicable groups [I], [Ho82]. Bridson and Tweedale also consider cases of more than two free factors.

7.5 Flower chains and Splitting with respect to field coefficients

The Reidemeister-Fox-matrix A with entries in $\mathbb{Z}(G * H)$ may also be considered for a partial semisplit presentation $\mathcal{P}$. Moreover, we may reduce the

$\mathbb{Z}$-coefficients of its entries mod some prime p or embed $\mathbb{Z}$ into the rationals. Furthermore, tensoring over prime fields, A becomes a matrix over the group ring $\mathbf{F}(G*H)$ with coefficients in a field $\mathbf{F}$; we continue to name it A. This matrix has *split rows* - those of shape $(*,\ldots,* \mid 0,\ldots,0)$ or $(0,\ldots,0 \mid *,\ldots,*)$ - and unsplit rows.

Define *elementary row operations* on such a matrix to be:

(I) Multiply some row by a nonzero element of $\mathbf{F}$, or add a $\mathbf{F}$-linear combination of some rows to another row.

(II) Add a $\mathbf{F}G$-linear combination of the split rows $(*,\ldots,* \mid 0,\ldots,0)$ to another split row of the same type $(*,\ldots,* \mid 0,\ldots,0)$; similarly with the roles of G and H interchanged.

(III) Discard zero rows and rows that are linear combinations of the others.

(IV) Permute rows.

Note that

- These row operations do not change the $\mathbf{F}(G*H)$-span of A.
- The number of rows does not increase under row operations.
- Despite the fact that $u_i \neq 1 \neq v_i$, $\widetilde{u}_i$ or $\widetilde{v}_i$ (or both) could be the zero-chain.

Throughout this section, all matrix entries are supposed to have field coefficients.

With the above operations we get in the case of a semisplit presentation $\mathcal{P}$ of $G*H$:

Theorem 7.4 *Let A be the Reidemeister-Fox-matrix of a semisplit presentation $\mathcal{P}$ for the free product $G*H$. Then there is a sequence of elementary row operations splitting A over $\mathbf{F}(G*H)$.*

Proof: All entries of matrices under consideration are assumed to have field coefficients. As in the Flower Theorem 7.3, construct a presentation $Q = \langle a_*, b_* \mid R_*(a_*), S_*(b_*), W_1, \ldots, W_m \rangle$ of $G*H$ with corresponding Reidemeister-Fox-matrix B. The rows for $\widetilde{R}_*$ and $\widetilde{S}_*$ already coincide for A and B and shall remain unchanged throughout. Like in the Steinitz exchange lemma for finite dimensional vector spaces we would like to trade the $\widetilde{T}_*$-rows of A for rows of B by performing elementary row operations on A and B.

To this end, we first transform B into some normal form using elementary operations of type (I) and (II) from above to below and, by (III), tacitly omitting all zero rows that arise at any time during the process (operation (IV) shall be used only for the Reidemeister-Fox matrix A).

Define matrices B^u and B^v as follows: If W_i is a word in $F(a_*)$ then $\widetilde{W}_i$ is a row vector of B^u and its twin chain is a row vector of B^v; if W_i is a word in $F(b_*)$ then $\widetilde{W}_i$ is a row vector of B^v and its twin chain is a row vector of B^u. Thus the rows of B^u have entries in $\mathbf{F}G$ and are the twins of the rows of B^v which have entries in $\mathbf{F}H$; in particular, B^u juxtaposed to B^v would make the size of the matrix containing the rows below the ones for $\widetilde{R}_*$ and $\widetilde{S}_*$ of B.

In B (and thus in B^u and in B^v as well) we tacitly omit all zero rows that might arise at any time during the process. Working from above to below, to B^u we perform the standard procedure to transform the $\mathbf{F}$ coefficients of the $\widetilde{u}_*$ into upper triangular form:

Assume, up to renumbering the T_* (and thus the u_* and v_* as well), that $\widetilde{u}_1$ has nonzero coefficient q_{11} in the first row of B^u, and multiply the corresponding rows of B, B^u and B^v by q_{11}^{-1}. Then add appropriate rational multiples of this row to the subsequent rows of B^u to eliminate their $\widetilde{u}_1$-summands; follow the procedure for B^v and perform the corresponding operations to B itself.

We claim that neither is the span of B changed nor the property of the $\widetilde{W}_*$ being flower chains: The twin chain of each $\widetilde{W}_i$-row is a $\mathbf{F}G$- resp. $\mathbf{F}H$- linear combination of the rows above.

1st case: If $\widetilde{W}_1$ and $\widetilde{W}_2$ lie in the same factor, so these rows coincide for B and B^u, say, use an elementary operation of type I for B and B^u and an elementary operation of type II for B^v. The second row of B^v remains a $\mathbf{F}H$-linear combination of the $\widetilde{S}_*$ and the first row.

2nd case: If $\widetilde{W}_1$ and $\widetilde{W}_2$ lie in different factors, the twin of the row $\widetilde{W}_1$ is a linear combination of the rows above and can thus be used for the desired elementary operation of type II on the second row. $\widetilde{W}_1$ itself is used to transform the twin of $\widetilde{W}_2$ accordingly by an elementary operation of type I. Thus the twin of the new $\widetilde{W}_2$-row of B is still a $\mathbf{F}G$- resp. $\mathbf{F}H$-linear combination of the rows above.

Continue until upper triangular form is achieved for B^u and B^v.

In particular we see that at this point B has at most n (nonzero) rows where n is the number of T_* in $\mathcal{P}$, $n \geq m$.

Now it is easy to substitute the $\widetilde{T}_*$-rows in A by corresponding rows of B: The row of B corresponding to $\widetilde{W}_1$ reads, say, $\left(\widetilde{u}_1 + \sum_{i>1} q_i \widetilde{u}_i \mid 0, \ldots, 0\right)$.

Add $\sum_{i>1} q_i \widetilde{T}_i$ to $\widetilde{T}_1 = (\widetilde{u}_1 \mid \widetilde{v}_1)$.

By the property of flower chains there is a linear combination of the $\widetilde{S}_*$

annihilating the right part of this row, i.e. transforming it into row $\widetilde{W}_1$ of B. Proceed with such replacements until all B-rows appear in A. If there is still any $\widetilde{T}_*$-row left, recall that Q is a presentation for $G*H$, hence all $(\widetilde{u}_* \mid 0, \ldots, 0)$ and $(0, \ldots, 0 \mid \widetilde{v}_*)$ are linear combinations of the rows of B. Thus those remaining rows of A could be discarded by means of elementary operations of type (III).
□

Note that a direct sequence of splitting operations according to Theorem 7.4 does not necessarily lead to the minimal number of relators for the factors.

What happens in the case of a partial semisplit presentation? Then the algebraic steps in the proof of Theorem 7.4 still result in a well-defined final span of the split vectors for both sides and a well-defined number $n_\mathbf{F}(\mathcal{P})$ of unsplit vectors.

(5) If this number $n_\mathbf{F}(\mathcal{P})$ is different from 0 for some field $\mathbf{F}$, then $\mathcal{P}$ isn't a complete semisplit presentation of $G * H$.

It may be an interesting project to investigate the situation of Theorem 7.4 and (5) for the original integral matrices A, B. One has to consider certain unimodular transformations taking care of elementary divisors.

Theorem 7.4 immediately implies

Corollary 7.5 *Let* $\mathcal{P} = \langle a_*, b_* \mid R_*, S_*, u_* v_*^{-1} \rangle$ *be a semisplit presentation of* $G * H$ *with Reidemeister-Fox-matrix A. Then*

$$\mathbf{N}_P \otimes_{\mathbb{Z}(G*H)} \mathbf{F}(G * H) \approx (\mathbf{N}_{\langle a_* \mid R_*, u_* \rangle} \otimes_{\mathbb{Z}G} \mathbf{F}(G * H)) \oplus (\mathbf{N}_{\langle b_* \mid S_*, v_* \rangle} \otimes_{\mathbb{Z}H} \mathbf{F}(G * H)).$$

Let us finally come to the notion of a "rational relation gap" which describes the situation that the minimal number of generators for the relation module $\mathbf{N}_\mathcal{P}$ with rational coefficients is smaller than the minimal number of relators which are needed to present a group π as $\pi = F(a_1, \ldots, a_g)/N$.

Assume $\mathcal{P}$ is a presentation of a free product with a rational relation gap; i.e. $\mathcal{P}$ has minimal number of relators, but taking Fox derivatives, one row can be eliminated. Then we obtain

Corollary 7.6 *If some semisplit presentation of a free product has a rational relation gap, then some presentation of one of the factors has also.*

By our calculations in Section 2 we know already that the groups $\mathbb{Z}_n \times \mathbb{Z}_n$

need only two generators for their rational relation modules and the Epstein groups ($\mathbb{Z}_k \times \mathbb{Z}$) only one. In particular these groups have a rational relation gap. This is consistent with Corollary 7.6.

7.6 Proof of the Theorem on semisplit Q^{**}-transformations

In [I], Chapter XII, section 3.2 we gave a list of *semisplit Q^{**}-transformations* which preserve the property of being a finite semisplit presentation of $\pi = G * H$. We repeat the theorem: (Theorem 3.4 of [I], Chapter XII):

Theorem 7.7 *Let $\mathcal{P}, \mathcal{Q}$ be finite semisplit presentations of $\pi = G * H$ and let $\mathcal{P} \to \mathcal{Q}$ be a Q^{**}-transformation which induces the identity map on π. Then there exists a semisplit Q^{**}-transformation from $\mathcal{P}$ to $\mathcal{Q}$.*

The condensed proof given here is based on the original idea of talks by W. Metzler and adapted from the Master thesis of Janina Körner [Kör15]:

(6) One encircles regions with 1-coloured boundary by 2-coloured ones and thus gets new ones of the other colour.

This is the same idea as for the Flower Theorem 7.3 of this chapter. A different approach due to the first author and S. Rosebrock was carried out in detail by Monika Eufinger [Euf92].

Proof: Let $\mathcal{P}$ and $\mathcal{Q}$ and a finite sequence of Q^{**}-transformations $\mathcal{P} = \mathcal{P}_1 \to \ldots \to \mathcal{P}_n = \mathcal{Q}$ be given with intermediate stages $\mathcal{P}_i$ which may be non semiplit.

We assume by [I], page 373, Theorem 3.1 that inductively up to a certain $\mathcal{P}_{i_0}$ the 2-cells have been subdivided into semiplit presentations such that all transformations are semiplit and the 2-cells have a *preferred 1-coloured 2-cell* in their subdivision.

For the next step we subdivide the 2-cells of $\mathcal{P}_{i_0+1}$, also with such 1-coloured preferred 2-cells. By the technique of STALLINGS' proof of the GRUŞKO-NEUMANN theorem we may assume that the 1-skeleton of the subdivided $\mathcal{P}_{i_0}$ is transformed to the one of $\mathcal{P}_{i_0+1}$ by semisplit generator transformations. Thus we have to deal with an elementary Q-transformation $\mathcal{P}_{i_0} \to \mathcal{P}_{i_0+1}$.

It is helpful to consider cyclic words and not to perform reductions in order to avoid squeezing off 2-dimensional material, as we have to keep track of which 2-cells are adjacent to each other. Nevertheless we call a 2-cell or its boundary 1-coloured resp. 2-coloured as if a (cyclic) reduction was performed.

As the inversion of a relator doesn't change a 2-complex, and conjugation by our convention preserves the property of being semisplit, we have to consider the cases when $\mathcal{P}_{i_0} \to \mathcal{P}_{i_0+1}$ consists of α) an ordinary multiplication of a relation to another from one side or β) of a conjugation.

In the case α), we have a cyclic word R and a subdivision of the corresponding 2-cell with a preferred 1-coloured 2-cell and the same data for another relator S. The multiplication which produces $R \cdot S$ is carried out by opening S, inserting R and dismissing with R. We now glue in a 2-cell with boundary $R \cdot S$, a semisplit subdivision consistent with the one on $R \cdot S$ and a preferred 1-coloured 2-cell in it. If we remove the interior of the latter, its boundary constitutes a *van Kampen-diagram* together with the subdivided R, S and the remainder of the subdivided $R \cdot S$. All 2-cells are contained only once in this diagram. The diagram has a flower structure such that in each generation (6) holds; and the subdivided $\mathcal{P}_{i_0}$ contains the original preferred relators as "centres" with the boundary words of the petals. According to the generations of the flower structure, one now has to inductively erase the relations for the centres from the presentation and add the ones of the outer boundary. If one hits the generation with the preferred 1-coloured 2-cell of R, this has to be erased and the previous steps have to be inverted. These are all semisplit Q^{**}-transformations. The new outer boundary has replaced the one of the preferred 2-cell. On the 2-sphere given by R, S and $R \cdot S$ it has been homotoped closer to the preferred 2-cell boundary of $R \cdot S$. Continuing this process, the preferred 1-coloured 2-cell of R will be removed and the one of $R \cdot S$ has been added. The case β) is a degenerate one, inserting a conjugating word into S which cyclically cancels without having an R for it.

Now collapse the 2-cell which "had to go" from the boundary of its preferred 1-coloured centre. The semisplit transition for the subdivided $\mathcal{P}_{i_0} \to \mathcal{P}_{i_0+1}$ has been achieved. In the case $i_0 + 1 = n$, where $\mathcal{P}_n = \mathcal{Q}$, one does so for all original 2-cells.

Finally it is possible to reduce the boundary words of the intermediate stages and their semisplit Q^{**}-transitions. □

Remark: There are examples in which it is necessary to start the above process with a lowest generation from the subdivided S that changes and later is reconstructed again. This phenomenon is responsible for the fact why the idea of proof doesn't automatically yield a relative version of Theorem 7.7, where relators which appear in the beginning and in the end would be expected to stay fixed throughout the chain of transformations.

References

[I] Hog-Angeloni, C., Metzler, W., and Sieradski, A. (eds). 1993. *Two dimensional Homotopy and Combinatorial Group Theory*. London Math. Soc. Lecture Note Series 197, Cambridge University Press.

[AKN09] A. K. Naimzada, S. Stefani, A. Torriero. 2009. *Networks, Topology and Dynamics*. Theory and applications to economics and social systems edn. Springer-Verlag. [Chapter 5].

[Ans91] Anshel, I. L. 1991. *A Freiheitssatz for a class of two-relator groups*. Journal of Pure and Applied Algebra, **72**, 207–250. [Chapter 7].

[Art81] Artamonov, V. A. 1981. *Projective, nonfree modules over group rings of solvable groups*. Math. USSR Sbornik., **116**, 232–244. [Chapter 1].

[BaMi16] Barmak, J., and Minian, E. 2016. *A new test for asphericity and diagrammatic reducibility of group presentations.* preprint, arXiv:1601.00604. [Chapter 4].

[BaGr69] Barnett, D., and Gruenbaum, B. 1969. *On Steinitz's theorem concerning convex 3-polytopes and on some properties of planar graphs*. Many Facets of Graph Theory, Proc. Conf. Western Michigan Uni., Kalamazoo/Mi. 1968, 27–40. [Chapter 5].

[Bau74] Baumslag, G. 1974. *Finitely presented metabelian groups*. Proc. Second international Conference in Group Theory, Lecture Notes in Math., **372**, 65–74. [Chapters 1, 6].

[BaSo62] Baumslag, G., and Solitar, S. 1962. *Some two-generator one-relator non-Hopfian groups*. Bull. Amer. Math. Soc., **68**, 199–201. [Chapter 1].

[BeWe99] Becker, T., and Weispfenning, V. 1999. *Groebner bases. A computational approach to commutative algebra*. Graduate Texts in Math., vol. 141. Springer-Verlag, New York. [Chapter 5].

[Ber] Berge, J. *SnapPea*. [Chapter 5].

[Ber78] Bergman, G. M. 1978. *The diamond lemma for ring theory*. Adv. in Math., **29 (2)**, 178–218. [Chapter 5].

[BeHi08]   Berrick, A., and Hillman, J. 2008. *The Whitehead Conjecture and $L^{(2)}$-Betti numbers*. In: Chatterji, I. (ed), *Guido's Book of Conjectures*. Monographie No. 40 De L' Enseignement Mathematique. [Chapter 4].

[BeDu79]   Berridge, P. H., and Dunwoody, M. J. 1979. *Non-free projective modules for torsion-free groups*. J. London Math. Soc., **19(2)**, 433–436. [Chapter 1].

[BeBr97]   Bestvina, M., and Brady, N. 1997. *Morse theory and finiteness properties of groups*. Invent. Math., **129**, 445–470. [Chapters 1, 4, 6, 7].

[BeLaWa97]   Beyl, F. R., Latiolais, M. P., and Waller, N. 1997. *Classification of 2-complexes whose finite fundamental group is that of a 3-manifold*. Proc. Edinburgh Math. Soc., **40**, 69–84. [Chapter 1].

[BeWa05]   Beyl, F. R., and Waller, N. 2005. *A stably-free nonfree module and its relevance for homotopy classification, case Q_{28}*. Algebraic & Geometric Topology, **5**, 899–910. [Chapter 1].

[BeWa08]   Beyl, F. R., and Waller, N. 2008. *Examples of exotic free 2-complexes and stably free nonfree modules for quaternion groups*. Algebraic & Geometric Topology, **8**, 1–17. [Chapter 1].

[BeWa13]   Beyl, F. R., and Waller, N. 2013. *The geometric realization problem for algebraic 2-complexes*. Preliminary version, unpublished, [Chapter 1].

[Big93]   Biggs, N. 1993. *Algebraic Graph Theory*. Cambridge mathematical library (2nd ed.) edn. Cambridge University Press. [Chapter 5].

[Bir13]   Biroth, L. 2013. *Heegaard-Diagramme der 3-Sphäre und die Andrews-Curtis-Vermutung*. Masterthesis, Johannes Gutenberg University Mainz 2013 (unpublished). [Chapter 2].

[BiSt80]   Bieri, R., and Strebel, R. 1980. *Valuations and finitely presented metabelian groups*. Proc. London Math. Soc. (3), **41**, 439–464. [Chapters 1, 6].

[Bla10]   Blaavand, Jakob. 2010. *3-manifolds derived from link invariants*. lecture notes, University of California, Berkeley. [Chapter 3].

[BlTu06a]   Blanchet, C., and Turaev, V. 2006a. *Axiomatic approach to Topological Quantum Field theory*. Elsevier Ltd. [Chapter 3].

[BlTu06b]   Blanchet, C., and Turaev, V. 2006b. *Quantum 3-Manifold Invariants*. Elsevier Ltd. [Chapter 3].

[Bob00]   Bobtcheva, Ivelina. 2000. *On Quinn's Invariants of 2-dimensional CW-complexes*. arXiv math. GT. [Chapter 3].

[BoQu05]   Bobtcheva, Ivelina, and Quinn, Frank. 2005. *The reduction of quantum invariants of 4-thickenings*. Fund. Math., **188**, 21–43. [Chapter 3].

[BoLuMy05]   Borovik, A.V., Lubotzky, A., and Myasnikov, A.G. 2005. *The finitary Andrews-Curtis conjecture*. In: Infinite groups: geometric, combinatorial and dynamical aspects. Progr. Math., vol. 248, pp.15-30, Birkhäuser, Basel. [Chapters 2, 3].

[Bri15]   Bridson, M. 2015. *The complexity of balanced presentations and the Andrews-Curtis conjecture*. arXiv:1504.04187. [Chapter 2].

References

[BrTw07] Bridson, M., and Tweedale, M. 2007. *Deficieny and abelianized deficiency of some virtually free groups.* Math. Proc. Cambridge Phil. Soc., **143**, 257–264. [Chapters 6, 7].

[BrTw08] Bridson, M. R., and Tweedale, M. 2008. *Putative relation gaps.* Guido's Book of Conjectures. Monographie No. 40 De L'Enseignement Mathématique. [Chapter 1].

[BrTw14] Bridson, M. R., and Tweedale, M. 2014. *Constructing presentations of subgroups of right-angled Artin groups.* Geom. Dedicata, **169**, 1–14. [Chapters 1, 6].

[Bro87] Brown, K. S. 1987. *Finiteness properties of groups.* J. Pure Appl. Algebra, **44**, 45–75. [Chapters 1, 6].

[Bro76] Browning, W.J. 1976. *Normal generators of finite groups.* manuscript. http://www.cambridge.org/9781316600900, [Chapter 2].

[Bur01] Burdon, M. 2001. *Embedding 2-polyhedra with regular neighborhoods which have sphere boundaries.* PhD thesis, Portland State University. [Chapter 5].

[CaEi56] Cartan, H., and Eilenberg, S. 1956. *Homological algebra.* Princeton Mathematical Series, vol. 19. Princeton, NJ: Princeton University Press. [Chapter 1].

[ChdW14] Christmann, M., and de Wolff, T. 2014. *A sharp upper bound for the complexity of labeled oriented trees.* preprint, arXiv:1412.7257. [Chapter 4].

[Coh64] Cohn, P. M. 1964. *Free ideal rings.* J. Algebra, **1**, 47–69. [Chapter 1].

[CoGrKo74] Cossey, J., Gruenberg, K. W., and Kovacs, L. G. 1974. *Presentation rank of a direct product of finite groups.* J. Algebra, **28**, 597–603. [Chapters 1, 6].

[Dun72] Dunwoody, M. J. 1972. *Relation modules.* Bull. London Math. Soc., **4**, 151–155. [Chapters 1, 6].

[Eck00] Eckmann, B. 2000. *Introduction to L^2-methods in Topology.* Israel J. Math., **117**, 183–219. [Chapter 4].

[EiGa57] Eilenberg, S. and Ganea, T. 1957. *On the Lusternik-Schnirelmann category of abstract groups.* Ann. of Math., **65**, 517–518. [Chapter 1].

[Eps61] Epstein, D. B. A. 1961. *Finite presentations of groups and 3-manifolds.* Quart. J. Math. Oxford Series, **(2), 12**. [Chapters 6, 7].

[Euf92] Eufinger, M. 1992. *Normalformen für Q^{**}-Transformationen bei Präsentationen freier Produkte.* Diplomarbeit, Frankfurt/Main (unpublished). [Chapter 7].

[Fox52] Fox, R. H. 1952. *On the Complementary Domains of a Certain Pair of Inequivalent Knots.* Indag. Math, **14**, 37–40. [Chapter 5].

[FrYe89] Freyd, Peter, and Yetter, David. 1989. *Brided compact closed Categories with applications to Low dimensional Topology.* Advances in Mathematics, **77**, 156–182. [Chapter 3].

[Geo08] Geoghegan, R. 2008. *Topological Methods in Group Theory.* Graduate Texts in Math., vol. 243. Springer-Verlag. [Chapters 1, 6].

References

[Gil76]   Gildenhuys, D. 1976. *Classification of solvable groups of cohomological dimension 2*. Math. Z., **166**, 21–25. [Chapters 1, 6].

[GlHo05] Glock, J., and Hog-Angeloni, C. I. 2005. *Embeddings of 2-complexes into 3-manifolds*. Journal of Knot Theory and Its Ramifications, **14 (1)**, 9–20. [Chapter 5].

[Gol99]   Goldstein, R. Z. 1999. *The length and thickness of words in a free group*. Proc. of the Am. Math. Soc., **127 (10)**, 2857–2863. [Chapter 5].

[Gru76]   Gruenberg, K. W. 1976. *Relation Modules of Finite Groups*. CBMS Regional Conference Series in Mathematics No. 25, AMS. [Chapters 1, 6].

[Gru80]   Gruenberg, K. W. 1980. *The partial Euler characteristic of the direct powers of a finite group*. Arch. Math., **35**, 267–274. [Chapters 1, 6].

[GrLi08] Gruenberg†, K., and Linnell, P. 2008. *Generation gaps and abelianized defects of free products*. J. Group Theory, **11 (5)**, 587–608. [Chapters 1, 6, 7].

[Guo16]   Guo, Guangyuan. 2016. *Heegaard diagrams of S^3 and the Andrews-Curtis Conjecture*. arXiv:1601.06871. [Chapter 2].

[Har93]   Harlander, J. 1993. *Solvable groups with cyclic relation module*. J. Pure Appl. Algebra, **190**, 189–198. [Chapter 1].

[Har96]   Harlander, J. 1996. *Closing the relation gap by direct product stabilization*. J. Algebra, **182**, 511–521. [Chapters 1, 6].

[Har97]   Harlander, J. 1997. *Embeddings into efficient groups*. Proc. Edinburgh Math. Soc., **40**, 314–324. [Chapters 1, 6].

[Har00]   Harlander, J. 2000. *Some aspects of efficiency*. Pages 165–180 of: Baik, Johnson, and Kim (eds), *Groups–Korea 1998, Proceedings of the 4th international conference, Pusan, Korea*. Walter deGruyter. [Chapters 1, 6].

[HaHoMeRo00] Harlander, J., Hog-Angeloni, C., Metzler, W., and Rosebrock, S. 2000. *Problems in Low-dimensional Topology*. Encyclopedia of Mathematics Supplement II (ed. M. Hazewinkel). Kluwer Academic Publishers. [Chapter 1].

[HaJe06] Harlander, J., and Jensen, J. A. 2006. *Exotic relation modules and homotopy types for certain 1-relator groups*. Algebr. Geom. Topol., **6**, 2163–2173. [Chapter 1].

[HaMi10] Harlander, J., and Misseldine, A. 2010. *On the K-theory and homotopy theory of the Klein bottle group*. Homology, Homotopy, and Applications, **12(2)**, 1–10. [Chapter 1].

[HaRo03] Harlander, J., and Rosebrock, S. 2003. *Generalized knot complements and some aspherical ribbon disc complements*. Knot theory and its Ramifications, **12 (7)**, 947–962. [Chapter 4].

[HaRo10] Harlander, J., and Rosebrock, S. 2010. *On distinguishing virtual knot groups from knot groups*. Journal of Knot Theory and its Ramifications, **19 (5)**, 695–704. [Chapter 4].

[HaRo12] Harlander, J., and Rosebrock, S. 2012. *On Primeness of Labeled Oriented Trees*. Knot theory and its Ramifications, **21 (8)**. [Chapter 4].

References

[HaRo15] Harlander, J., and Rosebrock, S. 2015. *Aspherical Word Labeled Oriented Graphs and cyclically presented groups.* Knot theory and its Ramifications, **24 (5)**. [Chapter 4].

[HaRo17] Harlander, J., and Rosebrock, S. 2017. *Injective labeled oriented trees are aspherical.* arXiv:1212.1943, to appear in: Mathematische Zeitschrift. [Chapter 4].

[Hig51] Higman, G. 1951. *A finitely generated infinite simple group.* J. London Math. Soc., **26**, 61–64. [Chapter 1].

[Hil97] Hillman, J. A. 1997. L^2-*homology and asphericity.* Israel J. Math., **99**, 271–283. [Chapter 4].

[Hän05] Hänsel, J. 2005. *Andrews-Curtis-Graphen endlicher Gruppen.* Diplomarbeit, Frankfurt (unpublished). [Chapter 2].

[Ho-AnMa08] Hog-Angeloni, C., and Matveev, S. 2008. *Roots in 3-manifold topology.* Pages 295–319 of: The Zieschang Gedenkschrift, Geom. Topol. Monogr., vol. 14. Geometry and Topology Publications, Coventry. [Chapter 5].

[Ho-AnMe04] Hog-Angeloni, C., and Metzler, W. 2004. *Ein Überblick über Resultate und Aktivitäten zum Andrews-Curtis-Problem.* preprint, Frankfurt. [Chapter 2].

[Ho-AnMe06] Hog-Angeloni, C., and Metzler, W. 2006. *Strategies towards a disproof of the general Andrews-Curtis Conjecture.* Siberian Electronic Mathematical Reports, **3**. [Chapter 2].

[How99] Howie, J. 1999. *Bestvina-Brady Groups and the Plus Construction.* Math. Proc. Cambridge Phil. Soc., **127**, 487–493. [Chapters 1, 4, 6].

[Hu01] Hu, Sen. 2001. *Lecture notes on Chern-Simons-Witten Theory.* World Scientific (Wspc). [Chapter 3].

[HuRo95] Huck, G., and Rosebrock, S. 1995. *Weight tests and hyperbolic groups.* Pages 174–183 of: A. Duncan, N. Gilbert, J. Howie (ed), *Combinatorial and Geometric Group Theory.* London Math. Soc. Lecture Note Ser., vol. 204. London: Cambridge University Press. [Chapter 4].

[HuRo00] Huck, G., and Rosebrock, S. 2000. *Cancellation Diagrams with nonpositive Curvature.* Pages 128–149 of: et al., Michael Atkinson (ed), *Computational and Geometric Aspects of Modern Algebra.* London Math. Soc. Lecture Note Ser., vol. 275. London: Cambridge University Press. [Chapter 4].

[HuRo01] Huck, G., and Rosebrock, S. 2001. *Aspherical Labelled Oriented Trees and Knots.* Proceedings of the Edinburgh Math. Soc. 44, 285–294. [Chapter 4].

[HuRo07] Huck, G., and Rosebrock, S. 2007. *Spherical Diagrams and Labelled Oriented Trees.* Proceedings of the Edinburgh Math. Soc., **137A**, 519–530. [Chapter 4].

[Joh97] Johnson, D.L. 1997. *Presentations of Groups, 2nd edition.* LMS Student Texts, vol. 15. Cambridge University Press. [Chapters 1, 6].

[Joh03] Johnson, F.E.A. 2003. *Stable Modules and the D(2)-Problem.* London Math. Soc. Lecture Note Ser., vol. 301. Cambridge University Press. [Chapter 1].

References

[Joh12] Johnson, F.E.A. 2012. *Syzygies and Homotopy Theory*. Algebra and Applications, vol. 17. London: Springer-Verlag. [Chapter 1].

[Kad10] Kaden, H. 2010. *Considerations about the Andrews-Curtis invariants based on sliced 2-complexes*. arXiv math. GT. [Chapter 3].

[Kad17] Kaden, H. 2017. *Considerations for constructing Andrews-Curtis invariants of s-move 3-cells*. arXiv math. GT. [Chapter 3].

[KaMaSo60] Karrass, A., Magnus, W., and Solitar, D. 1960. *Elements of finite order in groups with a single defining relation*. Comm. Pure Appl. Math., **13**, 57–66. [Chapter 7].

[KaRo96] Kaselowsky, A., and Rosebrock, S. 1996. *On the Impossibility of a Generalization of the HOMFLY – Polynomial to labelled Oriented Graphs*. Annales de la Faculté des Sciences de Toulouse V, **3**, 407–419. [Chapter 4].

[Kau87] Kauffman, Louis H. 1987. *On knots*. Annals of Mathematics Studies, vol. 115. Princeton University Press, Princeton, NJ. [Chapter 3].

[Kau99] Kauffman, Louis H. 1999. *Virtual knot theory*. European Journal of Combinatorics, **20 (7)**, 663–690. [Chapters 4, 5].

[Kin07a] King, Simon A. 2007a. *Ideal Turaev-Viro invariants*. Topology Appl., **154**(6), 1141–1156. [Chapter 3].

[Kin07b] King, Simon A. 2007b. *Verschiedene Anwendungen kombinatorischer und algebraischer Strukturen in der Topologie*. [Chapter 3].

[Kly93] Klyachko, A. 1993. *A funny property of sphere and equations over groups*. Communications in Algebra, **21 (7)**, 2555 – 2575. [Chapter 4].

[Kne29] Kneser, H. 1929. *Geschlossene Flächen in dreidimensionalen Mannigfaltigkeiten*. J. Dtsch. Math. Verein, **38**, 248–260. [Chapter 5].

[KoMa11] Korablev, F. G., and Matveev, S. V. 2011. *Reduction of knots in thickened surfaces and virtual knots*. Dokl. Math., **83 (2)**, 262–264. [Chapter 5].

[KrRo00] Kreuzer, Martin, and Robbiano, Lorenzo. 2000. *Computational commutative algebra. 1*. Springer-Verlag, Berlin. [Chapter 3].

[Kör15] Körner, J. 2015. *Das Relatorenlückenproblem für freie Produkte*. Masterarbeit, Frankfurt/Main, unpublished. [Chapter 7].

[Küh00a] Kühn, A. 2000a. extract from: *Stabile Teilkomplexe und Andrews-Curtis-Operationen*. Diplomarbeit, Frankfurt/Main. http://www.cambridge.org/9781316600900, [Chapter 2].

[Küh00b] Kühn, A. 2000b. *Stabile Teilkomplexe und Andrews-Curtis-Operationen*. Diplomarbeit, Frankfurt/Main (unpublished). [Chapter 2].

[Lüc01] Lück, W. 2001. L^2-*invariants: Theory and Applications to Geometry and K-Theory*. Ergebnisse der Mathematik und ihrer Grenzgebiete, Volume 44, Springer. [Chapter 4].

[Li00] Li, Zhongmou. 2000. *Heegaard-Diagrams and Applications*. PhD-thesis, university of British Columbia. [Chapter 2].

[Lis17] Lishak, B. 2017. *Balanced finite presentations of the trivial group*. Journal of Topology and Analysis, **9 (2)**, 363–378. [Chapter 2].

[LiNa17]	Lishak, B., and Nabutovsky, A. 2017. *Balanced presentations of the trivial group and four-dimensional geometry*. Journal of topology and analysis, **9 (1)**, 27–49. [Chapter 2].
[Lou15]	Louder, L. 2015. *Nielsen equivalence in closed surface groups*. arXiv:1009.0454v2. [Chapter 1].
[Luf96]	Luft, E. 1996. *On 2-dimensional aspherical complexes and a problem of J. H. C. Whitehead*. Math. Proc. Camb. Phil. Soc., **119**, 493–495. [Chapter 4].
[Lus95]	Lustig, M. 1995. *Non-efficient torsion free groups exist*. Comm. Algebra, **23**, 215–218. [Chapters 1, 6].
[MaKaSo76]	Magnus, W., Karras, A., and Solitar, D. 1976. *Combinatorial Group Theory*. Dover Publications. [Chapter 1].
[Man80]	Mandelbaum, Richard. 1980. *Four-dimensional topology: an introduction*. bams. [Chapter 3].
[Man07a]	Mannan, W. H. 2007a. *The $D(2)$ property for D_8*. Algebraic & Geometric Topology, **7**, 517–528. [Chapter 1].
[Man07b]	Mannan, W. H. 2007b. *Homotopy types of truncated projective resolutions*. Homology, Homotopy and Applications, **9(2)**, 445–449. [Chapter 1].
[Man09]	Mannan, W. H. 2009. *Realizing algebraic 2-complexes by cell complexes*. Math. Proc. Cambridge Math. Soc., Issue 03, **146, Issue 03**, 671–673. [Chapter 1].
[Man13]	Mannan, W. H. 2013. *A commutative version of the group ring*. J. Algebra, **379**, 113–143. [Chapter 1].
[MaO'S13]	Mannan, W. H., and O'Shea, S. 2013. *Minimal algebraic complexes over D_{4n}*. Algebraic & Geometric Topology, **13**, 3287–3304. [Chapter 1].
[Man16]	Mannan, Wajid H. 2016. *Explicit generators of the relation module in the example of Gruenberg-Linell*. Math. Proc. Cambridge Philos. Soc., **161 (2)**, 199–202. [Chapter 7].
[Mat03]	Matveev, S. 2003. *Algorithmic Topology and Classification of 3-Manifolds*. Algorithms and Computations in Mathematics, vol. 9. Springer Verlag New York, Heidelberg, Berlin. [Chapters 2, 3].
[Mat12a]	Matveev, S. 2012a. *Prime decomposition of knots in $T \times I$*. Topology Appl., **159 (7)**, 1820–1824. [Chapter 5].
[Mat12b]	Matveev, S. 2012b. *Roots and decompositions of three-dimensional manifolds*. Russiam Math. Surveis, **67 (3)**, 1459–507. [Chapter 5].
[MaTu11]	Matveev, S., and Turaev, V. 2011. *A semigroup of theta-curves in 3-manifolds*. Mosc. Math. J., **11 (4)**, 805–814. [Chapter 5].
[Mat10]	Matveev, S. V. 2010. *On prime decompositions of knotted graphs and orbifolds*. Atti Semin. Mat. Fis. Univ. Modena Reggio Emilia, **57**, 89–96. [Chapter 5].
[MaSo96]	Matveev, Sergei V., and Sokolov, Maxim V. 1996. *On a simple invariant of Turaev-Viro type*. Zap. Nauchn. Sem. S.-Peterburg. Otdel. Mat. Inst. Steklov. (POMI), **234** (Differ. Geom. Gruppy Li i Mekh. 15-1), 137–142, 263. [Chapter 3].

References

[Met00]   Metzler, W. 2000. *Verallgemeinerte Biasinvarianten und ihre Berechnung.* Pages 192–207 of: Atkinson, M. et al. (ed), *Computational and Geometric Aspects of Modern Algebra.* London Math. Soc. Lecture Note Series, vol. 275. [Chapter 2].

[Mil62]   Milnor, J. 1962. *A unique factorisation theorem for 3-manifolds.* Amer. J. Math., **84**, 1–7. [Chapter 5].

[Mül00]   Müller, Klaus. 2000. *Probleme des Einfachen Homotopietyps in niederen Dimensionen und ihre Behandlung mit Hilfsmitteln der Topologischen Quantenfeldtheorie.* Der Andere Verlag Dissertation Frankfurt/Main. [Chapter 3].

[Mut11]   Muth, C. 2011. *Relatorenlücke und vermutete Beispiele.* Masterarbeit, Mainz unpublished. [Chapter 7].

[Nab12]   Naber, Greg. 2012. *Yang-Mills to Seiberg-Witten via TQFT The Witten Conjecture.* preprint, Black Hills State University. [Chapter 3].

[New42]   Newman, M. H. A. 1942. *On theories with a combinatorial definition of 'equivalence'.* Ann. of Math., (2) **43:2**, 223–243. [Chapter 5].

[O'S12]   O'Shea, S. 2012. *The D(2)-problem for dihedral groups of order 4n.* Algebraic & Geometric Topology, **12**, 2287–2297. [Chapter 1].

[Osi15]   Osin, D. 2015. *On acylindrical hyperbolicity of groups with positive first ℓ^2-Betti number.* Bull. London Math. Soc. 47, **5**, 725–730. [Chapter 4].

[OsTh13] Osin, D., and Thom, A. 2013. *Normal generation and ℓ^2-betti numbers of groups.* Math. Ann., **355 (4)**, 1331–1347. [Chapters I, IV].

[Per02]   Perelman, G. 2002. *The entropy formula for the Ricci flow and its geometric applications.* arXiv:math.DG/0211159. [Chapter 2].

[Per03a] Perelman, G. 2003a. *Finite extinction time for the solutions to the Ricci flow on certain three-manifolds.* arXiv:math.DG/0307245. [Chapter 2].

[Per03b] Perelman, G. 2003b. *Ricci flow with surgery on three-manifolds.* arXiv:math.DG/0303109. [Chapter 2].

[Pet07]   Petronio, C. 2007. *Spherical splitting of 3-orbifolds.* Math. Proc. Cambridge Philos. Soc., **142**, 269–287. [Chapter 5].

[Poe13]   Poelstra, Andrew. 2013. *A brief Overview of Topological Quantum Field Theory.* preprint. [Chapter 3].

[Qui76]   Quillen, D. 1976. *Projective modules over polynomial rings.* Invent. Math., **36**, 167–171. [Chapter 1].

[Qui92]   Quinn, Frank. 1992. *Lectures on Axiomatic Quantum Field Theory.* preprint. [Chapter 3].

[Qui95]   Quinn, Frank. 1995. *Lectures on Axiomatic Quantum Field Theory.* IAS/Park City Mathematical series, **1**. [Chapter 3].

[Rep88]   Repovs, D. 1988. *Regular neighbourhoods of homotopically PL embedded compacta in 3-manifolds.* Suppl. Rend. Circ. Mat. Palermo, **18**, 213–243. [Chapter 5].

[Ros94]   Rosebrock, S. 1994. *On the Realization of Wirtinger Presentations as Knot Groups.* Journal of Knot Theory and its Ramifications, **3 (2)**, 211–222. [Chapter 4].

References 175

[Ros00] Rosebrock, S. 2000. *Some aspherical labeled oriented graphs*. Pages 307–314 of: Matveev, S. (ed), *Low-Dimensional Topology and Combinatorial Group Theory*. Proceedings of the International Conference, Kiev. [Chapter 4].

[Ros07] Rosebrock, S. 2007. *The Whitehead-Conjecture – an Overview*. Siberian Electronic Mathematical Reports, **4**, 440–449. [Chapter 4].

[Ros10] Rosebrock, S. 2010. *On the Complexity of labeled oriented trees*. Proc. of the Indian Acad. of Sci, **120** (1), 11–18. [Chapter 4].

[Ros02] Rosson, John. 2002. *Multiplicative Invariants of Special 2-Complexes*. Ph.D. thesis, Department of Mathematics, Portland State University, unpublished. [Chapter 3].

[Rot02] Rotman, J. J. 2002. *Advanced Modern Algebra*. Prentice Hall. [Chapters 1, 6].

[Sch49] Schubert, H. 1949. *Die eindeutige Zerlegbarkeit eines Knotens in Primknoten*. S.-B. Heidelberger Akad. Wiss. Math.-Nat. Kl., **3**, 57–104. [Chapter 5].

[Sta85] Stafford, J. T. 1985. *Stably free, projective right ideals*. Compositio Math., **54**, 63–78. [Chapter 1].

[Sta99] Stallings, J. 1999. *Whitehead Graphs on handlebodies*. Geom. group theory down under (Canberra 1996) edn. de Gruyter, Berlin. [Chapter 5]. Pages 317–330.

[Sta68] Stallings, J. R. 1968. *On torsion-free groups with infinitely many ends*. Ann. of Math., **88**, 312–334. [Chapter 1].

[Sta87] Stallings, John. 1987. *A graph-theoretic lemma and group-embeddings*. Pages 145–155 of: Gersten, S.M., and Stallings, J.R. (eds), *Combinatorial group theory and topology*. Annals of Mathematics Studies, vol. 111. [Chapter 4].

[Str74] Strebel, R. 1974. *Homological methods applied to the derived series of groups*. Comment. Math. Helv., **49**, 63–78. [Chapter 1].

[Swa60] Swan, R. G. 1960. *Periodic resolutions for finite groups*. Ann. of Math. (2), **72**, 267–291. [Chapter 1].

[Swa69] Swan, R. G. 1969. *Groups of cohomological dimension one*. J. Algebra, **12**, 585–601. [Chapter 1].

[Swa83] Swan, R. G. 1983. *Projective modules over binary polyhedral groups*. J. Reine Angew. Math., **342**, 66–172. [Chapter 1].

[Swa70] Swarup, G. A. 1970. *Some properties of 3-manifolds with boundary*. Quart. J. Math. Oxford Ser., (2) **21:1**, 1–23. [Chapter 5].

[ThSe30] Threlfall, W., and Seifert, H. 1930. *Topologische Untersuchung der Diskontinuitätsbereiche endlicher Bewegungsgruppen des dreidimensionalen sphärischen Raumes I*. Math. Ann., **104**, 1–70. [Chapter 1].

[ThSe32] Threlfall, W., and Seifert, H. 1932. *Topologische Untersuchung der Diskontinuitätsbereiche endlicher Bewegungsgruppen des dreidimensionalen sphärischen Raumes II*. Math. Ann., **107**, 543–586. [Chapter 1].

[TuVi92] Turaev, V., and Viro, O. 1992. *State Sum Invariants of 3-manifolds and Quantum 6j-symbols*. Topology, **31** (4), 865–902. [Chapter 3].

References

[Tur94] Turaev, V. G. 1994. *Quantum invariants of knots and 3-manifolds.* de Gruyter Studies in Mathematics, vol. 18. Walter de Gruyter & Co., Berlin. [Chapter 3].

[Tut61] Tutte, W. 1961. *A theory of 3-connected graphs.* Ind. Math., **29**, 441–455. [Chapter 5].

[Tut63] Tutte, W. 1963. *How to draw a graph.* Proc. London Math. Soc. (3), **13**, 743–768. [Chapter 5].

[Vir10] Virelizier, Alexis. 2010. *Quantum invariants of 3-manifolds, TQFTs, and Hopf monads.* Habilitation a Diriger des recherches Universite Montpellier 2. [Chapter 3].

[Wald68] Waldhausen, F. 1968. *Heegaard-Zerlegungen der 3-Sphäre.* Topology, **7**, 195–203. [Chapter 2].

[Wall66] Wall, C. T. C. 1966. *Finiteness conditions for CW-complexes II.* Proc. Roy. Soc. London Ser. A, **295**, 129–139. [Chapter 1].

[Wam70] Wamsley, J. W. 1970. *The multiplier of finite nilpotent groups.* Bull. Austral. Math. Soc. 3, 1–8. [Chapters 1, 6].

[Whi36a] Whitehead, J. H. C. 1936a. *On certain sets of elements in a free group.* Proc. London Math. Soc., **41**, 48–56. [Chapter 5].

[Whi36b] Whitehead, J. H. C. 1936b. *On equivalent sets of elements in a free group.* Annals of Math., **37**, 782–800. [Chapter 5].

[Whi39] Whitehead, J. H. C. 1939. *On the asphericity of regions in a 3-sphere.* Fund. Math., **32**, 149–166. [Chapter 4].

[Whi32a] Whitney, H. 1932a. *Congruent graphs and the connectivity of graphs.* Amer. J. Math., **54**, 150–168. [Chapter 5].

[Whi32b] Whitney, H. 1932b. *Non-separable and planar graphs.* Trans. Amer. Math. Soc., **34**, 339–362. [Chapter 5].

[Whi33a] Whitney, H. 1933a. *2-Isomorphic graphs.* Amer. J. Math., **55**, 245–254. [Chapter 5].

[Whi33b] Whitney, H. 1933b. *On the classification of graphs.* Amer. J. Math., **55**, 236–244. [Chapter 5].

[Whi33c] Whitney, H. 1933c. *A set of topological invariants for graphs.* Amer. J. Math., **55**, 231–235. [Chapter 5].

[Zen05] Zentner, Stefanie. 2005. *Wurzeln von Cobordismen und die Andrews-Curtis-Vermutung.* Diplomarbeit Frankfurt/Main (unpublished). [Chapter 5].

[Zie70] Zieschang, H. 1970. *Über die Nielsensche Kürzungsmethode in freien Produkten mit Amalgam.* Invent. Math., **10**, 4–37. [Chapter 7].

Index

$(G, 2)$-complex, 4
1-parameter family of graphs, 40
L^2 betti-number, 99
L^2-homology, 99
R-reduction, 50
T move, [I] and 61
T_3 move, [I] and 42
T_i moves, [I] and 42
U_i-stratum, 59
U_i-symbol, 64
$\mathcal{G}$-decorated, 60
$\mathcal{G}$-decoration, 59
$n_\mathbf{F}(\mathcal{P})$, 164
$\mathbb{Z}$-separated, 112
$\mathbf{N}_\mathcal{P}$, 156
3j-symbol, 65
4-term exact sequence, 3
6j-symbol, 65

acylindrical action, 101
acylindrically hyperbolic, 101
admissible colouring, 69
algebraic $(G, 2)$-complex, 4
algebraic 2-type, [I] and 13
almost presentation, 4
ambialgebra, 40
articulation point, 108
aspherical, [I] and 14

balanced, [I] and 147
bias invariant, [I] and 30
Bobtcheva's invariant, 48
boundary reducible LOG, 80
boundary strata, 59
boundary type, 59
branch points, 44
bubble move, 70

cancellation, 21
categories of TQFT, $\mathcal{M}$, $\partial \mathcal{M}$, $\partial^2 \mathcal{M}$,
    $Bord(\mathcal{M})$, 38, 39
Cayley graph, 3
circulator, 43, 46, 47
cohomological dimension, [I] and 9, 10
collar, 38
complexity, 80
compressed LOG, 80
corner
    in a TQFT, 39
    in a van Kampen-diagram, 75
corresponding strata, 60
coupon category, 48
cut vertex, 108
cycle of corners
    in a van Kampen-diagram, 75

D(2)-problem, 9
diagrammatically reducible, [I] and 74
dual object, 44
dual pairing, 52

E-group, 142
efficient, [I] and 129, 138
Eilenberg-Ganea conjecture, 10, 136
Epstein groups, 157
equivalent states, 64
Euler characteristic, [I] and 6

finite calculus, 61
flag complex, 134
Flower Theorem, 151
flower-relation, 151
Fox ideals, [I] and 139
free cancellation, 21
full subgraph, 75

generation gap, 128

Index

generation gap problem, 128
geometric realization problem, 1
geometrically realizable, 7
Gröbner bases, 67

height matrix, 78
Hilbert G-module, 98
Hilbert subspace, 98
homology equivalence, [I] and 30

I-test, 77
ideal state sum invariant, 67
incoming boundary, 37
injective LOT, 86
inner stratum, 60
interesting groups, 147
interior reducible, 80
isolated, 110

join, 134

labelled oriented circle, 97
labelled oriented graph, [I] and 73
labelled oriented interval, 97
labelled oriented tree, 74
long virtual knot diagram, 97
longitudinal identification, 55

minimal, 138
minimal LOT, 96
modular TQFT, 39
multiplicative property, 50

negative graph, 75
normal form, 67
normal rank, 101

outer stratum, 60
outgoing boundary, 37

pair of cut vertices, 108
partial semisplit presentation, 155
petal, 151
positive graph, 75
presentation, [I] and 2
proficient, 129
proper sub-LOG, 80

Q^{**}-transformation, [I] and 48, 150
Q-graph, 28
Q-transformation, [I] and 56, 160
quantum 6j-invariant, 68
Quinn invariant, 48

rational relation gap, 149
reduced LOG, 80
reduced Whitehead graph, 107
reducible spherical diagram, 74

reductions of modular invariants, 50
Reidemeister-Fox-derivative, [I] and 156
relation gap, 128, 129, 149
relation gap problem, 1, 35, 128
relation group, 2
relation lifting problem, 1
relation module, [I] and 2
relative Q^{**}-transformation, 30
relative Stallings test, 88
relatively vertex aspherical, 75
relatively vertex reduced, 75
reorientation of a LOT, 86
Reshetikhin–Turaev invariant, 36
right angled Artin group, 134

s-move, 53
semisimple tensor categories, 44
semisplit presentation, [I] and 150
separated, 109
separation points, 156
simple graph, 107
simple polyhedron, 60, 64
sliced 2-complex, 41
spacetime, 37
spacetime category TQFT, $\mathcal{M}$, 38
special ambialgebras, 46
special polyhedron, [I] and 42, 60
special spine, [I] and 60
spherical diagram, 74
spherical elements, 55
stably free cancellation, 21
standard 2-complex $K(P)$, [I] and 2
state, 64
state modules, 37
state sum, 64
sub-LOG, 80
surface group, 17
syllables of length 2, 107

$\mathcal{T}$, tensor category, 44
thickening, 105
Total collision, 81
TQFT, 37
trace, 98
trace unit, 46
Turaev–Viro invariant, 36, 68
type F_n, 134
type FP_n, 134
typical neighbourhood, 58

valency, 110
van Kampen-diagram, [I] and 166
vertex aspherical, 75

virtual link diagram, 96
von Neumann algebra, 98
von Neumann dimension, 99

Whitehead graph, 106
Whitehead's asphericity conjecture, [I] and 72, 136
WLOG-presentation, 82
word labelled oriented graph, 81
word labelled oriented tree, 82

Erratum

In [I] on page 44 the footnote was based on an incomplete chain of citations. As correspondence revealed, the question, whether the topological embeddability of a compact 2-polyhedron into Euclidean 4-space always implies the p.l. one, may even be an open problem.

Printed in the United States
by Baker & Taylor Publisher Services